SOUTHERN AFRICAN

TREES

A Photographic Guide

Piet van Wyk

STRUIK

Dedicated to my father, who introduced me to the
wonderful world around us when I was just a child.

Struik Publishers (Pty) Ltd
(a member of Struik New Holland Publishing (Pty) Ltd)
Cornelis Struik House
80 McKenzie Street
Cape Town 8001

Reg. No. 54/00965/07

First published in 1993
Second impression 1994
Third impression 1996
Fourth impression 1997
Fifth impression 1998

Text © Piet van Wyk 1993
Photographs © Piet van Wyk 1993
(except cover © Gerald Cubitt 1993;
elsewhere as indicated)
Maps © Piet van Wyk 1993
Illustrations © Nicci Page 1993

Edited by Nicola Marshall
Designed by René Greeff
DTP make-up by Suzanne Fortescue, Struik DTP
Reproduction by Hirt & Carter Cape (Pty) Ltd
Printing and binding by Tien Wah Press (Pte) Limited, Singapore

ISBN 1 86825 307 4

Cover photograph: *Colophospermum mopane* Mopane
Title page photograph: *Rothmannia capensis* Cape gardenia

Contents

Acknowledgements 4

Introduction 5

The vegetation of southern Africa 5

How to use this book 7

Further reading and references 7

Species accounts 8

Glossary 140

Index 142

Acknowledgements

Anybody with some experience of field studies will appreciate that the cost of compiling a publication of this nature is enormous, since it entails extensive travelling and the consumption of vast quantities of film. Without the very generous assistance, therefore, of the companies, institutions and persons mentioned below, this book would never have been possible. I salute them all, since they deserve the deepest gratitude from not only myself but also the eventual users of this book and of other related publications currently in preparation.

Companies and institutions
Mazda Wildlife Fund has placed a Mazda B2200 bakkie, with canopy, at my disposal.
Agfa SA (Pty) Ltd supplies and develops all my film.
Total SA (Pty) Ltd keeps the bakkie's fuel tank filled.
APBCO (Insurance brokers, Pretoria) covers the insurance of the Mazda bakkie.
Pick 'n Pay Stores Ltd.
The University of Pretoria and the *Rand Afrikaans University*, whose departments of botany assist me academically, financially and administratively.

Individuals
Official
Agfa SA: Ron Crane, Marthinus Bezuidenhout, Barbara Garner; APBCO Brokers: Kobus du Plessis; Mazda Wildlife Fund: Peter Frost; Pick 'n Pay Stores: Brenda van der Schijff; Rand Afrikaans University: Ben-Erik van Wyk; Total SA: Andries van der Walt; University of Pretoria: Braam van Wyk, Elsa van Wyk, Marthie Dettnam.

Private
All are friends – some new, some of long-standing – who have helped me and my wife Emmarentia by providing accommodation, hospitality, access to national parks, nature reserves and botanical gardens, or rendered highly valued assistance with fieldwork: Jonathan Gibson (Chobe Lodge, Botswana); Chris Burgers, Neil and Thea Fairall (CPA Nature Conservation); Wayne Matthews, Nick Steele, Harold and Marieta Thornhill (Kwazulu Bureau for Natural Resources); Lloyd and June Wilmot and staff (Lloyd's Camp, Botswana); Roelf van Wyk (my brother); Daan Botha, Kobus Eloff and Dawie Strydom (National Botanic Gardens); Robbie Robinson (National Parks Board); Harold and Tony Braack (Richtersveld National Park); Veronica Roodt (student, Botswana); Niek Hanekom, Andrew Spies (Tsitsikamma Coastal National Park); Sarel Yssel (West Coast National Park).

Introduction

A major factor that has influenced the growing interest in trees is the number of 'green' awareness campaigns that are being conducted around the world. As little as ten years ago, scientists who sounded warnings about global warming (the greenhouse effect), ozone depletion, the unacceptable rate at which rain forests are being cleared, acid rain, and so on, were either ignored or passed off as alarmists. Today the seriousness of pollution and its destructive effects on our environment are widely acknowledged, prompting people to engage in positive action to avert an impending catastrophe. One of the first steps we can take in this regard is to familiarize ourselves with an area of concern and then become thoroughly acquainted with the remedy and the ways in which it should be administered.

The purpose of this photographic guide is to 'whet the appetite' of the uninitiated and to hopefully encourage a desire among readers to become more knowledgeable about trees in general – and those of southern Africa in particular – and their critical role in the environment.

The vegetation of southern Africa

Based on floristic composition, six so-called Plant Kingdoms are recognized in the world. Two of them occur in southern Africa, namely, the Cape Fynbos Kingdom, which roughly covers the winter rainfall area of the Cape and the Palaeotropic Kingdom, which covers nearly all of Africa. Taking into account obvious differences in plantlife forms, as well as the composition and structure of plant communities, seven fairly marked vegetation zones (or biomes) can be distinguished in southern Africa. Based on plantlife forms and climate, they are: savanna (bushveld), forest, desert, succulent Karoo, Nama Karoo, grassland and fynbos.

The largest of these areas, the *savanna* biome (also referred to as bushveld or woodland), extends northwards from the eastern Cape to cover parts of Transkei, Natal and Swaziland, over 50 per cent of the Transvaal, the northern Cape and north-western Orange Free State, Mozambique, Zimbabwe, Botswana and northern Namibia. This region harbours most of the tree species in the subcontinent.

By far the smallest biome – *forest* – occurs intermittently and sometimes in only extremely small patches along the eastern escarpment in South Africa, from the southern Cape to the Soutpansberg in the northern Transvaal. Several hundred kilometres further north it reappears in the eastern, mountainous area of Zimbabwe and the western area of Mozambique.

Although insignificant in extent, the *fynbos* biome is floristically extremely rich and complex. The number of tree species, however, is fairly limited, albeit mostly unique.

The *desert* biome mainly corresponds with the Namib Desert in Namibia and extends in a relatively wide belt all along the coast northwards from Lüderitz. Tree species able to tolerate the harsh environmental conditions of this region are very limited.

Immediately south of the desert biome, and in a similar belt to the west of the western escarpment, the *succulent Karoo* biome extends southwards into South Africa, at first along the coast and then on the inland side of the fynbos biome to the Little Karoo. Tree species are more numerous in this biome but still limited. They include some succulents such as *Aloe* species.

5

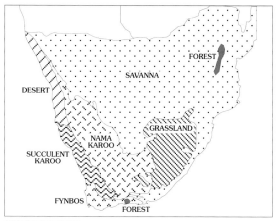

VEGETATION ZONES

The *Nama Karoo* biome encompasses the central plateau of the Cape Province, north of the southern mountain range, extending east-west, including Lesotho, the south-west Orange Free State, the southern interior of Namibia, and smaller parts in the eastern Cape. As can be expected, only tree species that are drought-resistant and frost-tolerant can survive in this biome, which therefore excludes the bulk of tree species in the subcontinent.

Despite higher rainfall, the very low winter temperatures have resulted in much the same situation in the last biome – the *grassland* biome. This area is wedged in-between the savanna biome on the southern, eastern, northern and north-western sides and the Nama-Karoo biome on the south-western side, therefore covering the eastern and southern Transvaal, the larger part of the Orange Free State, Lesotho, western Natal and sections of the eastern Cape.

Most of the tree species are therefore limited to two biomes, namely the forest and savanna areas of southern Africa. The reason for this is that the majority of plant species currently inhabiting the subcontinent started migrating southwards from the tropical region of Africa at the end of the Ice Age. Obviously, they were adapted to high temperatures, especially during winter, and favourable moisture conditions during summer, both of which are prevalent in the biomes involved.

Within each of the biomes conspicuous differences occur in the composition, structure and density of plant communities. These variations are attributable to the influence of moisture in an area, as well as differences in altitude, slope of the terrain, soil type and the prevalence of veld fires. In this regard soil is one of the major factors, which explains why the text frequently draws attention to the soil-preferences of the different species.

As only a fraction of the roughly 1 500 tree species in southern Africa could be included in this publication, drawing up a list proved to be a major task. Each tree was finally selected according to one or more of the following criteria: it must be widespread; have attractive or interesting flowers and/or fruits; represent an outstanding group of related species; be an important component in a specific biome; be a valuable source of plant or wood; have outstanding potential as a garden subject, or simply be impressive.

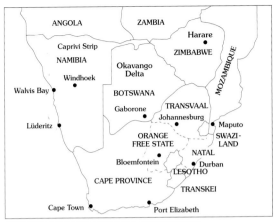

SOUTHERN AFRICAN PLACE NAMES AND LOCATIONS

How to use this book

When encountering a tree with which you are not familiar, establish whether it is dealt with in the book by comparing its bark,
leaves, flowers and/or fruit with the illustrations, making sure that
the distribution indicated on the map coincides with your own geo-
graphic location. If you think you have identified the tree, read
through the text to double check. Now look at the illustrations
again. Whether you have correctly identified your tree or not,
hopefully your interest (or frustration!) will at this stage have been
sufficiently aroused to make you hasten to the nearest bookshop to
buy a more comprehensive book.

Further reading and references

Coates Palgrave, K. 1983. *Trees of Southern Africa*. 2nd ed. C.
  Struik, Cape Town.
Coates Palgrave, K., P. & M. 1987. *Everyone's Guide to Trees of
  South Africa*. 2nd ed. Central News Agency, Johannesburg.
Moll, E. 1992. *Trees of Natal*. University of Cape Town Press,
  Cape Town.
Rourke, J.P. 1982. *The Proteas of Southern Africa*. Centaur Pub-
  lishers, Johannesburg.
Van Wyk, P. 1984. *Field Guide to the Trees of the Kruger Nation-
  al Park*. Struik Publishers, Cape Town.
Von Breitenbach, F. 1985. *Southern Cape Tree Guide*. Govern-
  ment Printer, Pretoria.
Von Breitenbach, F. & J. 1981. *National List of Indigenous Trees*.
  Dendrological Society, Pretoria.

Journals
Journal of Dendrology. Dendrological Society, Pretoria.

Information contained in articles written by various authors, mainly Von Breitenbach, was freely used during the compilation of this
publication.

Bark

Male cones

Podocarpus species (yellowwoods), especially *P. falca-tus* which *reaches a height of between 45 and 50 m,* are among the tallest trees in southern Africa. Tourists visiting the eastern Cape and travelling on the famous Garden Route, cannot fail to notice these trees

Berries

as the crowns of old trees tower above the forest canopy. Identification is further facilitated by the fact that most of them are 'decorated' with a light grey lichen (old man's beard) which hangs from the branches in long strings.

Everybody with an interest in trees or furniture is familiar with the fact that this tree, and its relative *P. latifolius*, occurs in the forests of the southern Cape, east and west of Knysna, which is the Mecca of yellowwood furniture in South Africa. Both trees also occur in all the evergreen forests along the escarpment up to the northern Transvaal, extending into Zimbabwe, Mozambique and the tropical areas to the north.

The trunk of the Outeniqua yellowwood can be up to 3 m in diameter, and is always long and straight, sometimes slightly fluted. The dark brown **bark** peels in large strips. The simple **leaves** are the smallest of the four species (up to 5 x 0,5 cm) and are straight or slightly falcate (sickle-shaped). The **fruit** is a round berry which becomes yellow at ripening. Berries are borne on female trees only. Male **cones** are catkin-like (scaly).

 This tree inhabits the escarpment mountains from Cape Town to the Natal midlands, the Drakensberg north of Nelspruit and the Soutpansberg, as well as areas in the Waterberg region of the Transvaal. It also extends into eastern Zimbabwe/Mozambique.

Male cones

Evergreen with a single trunk and a spreading, fairly dense crown, it usually attains a height of about 6 m. The **bark** on old stems is grey and flakes in long, narrow strips, revealing the reddish-brown living bark. *Young **leaves** are needle-like and old ones scale-like, minute and dark green.*

*Male and female **cones** are borne on separate trees.* Male cones are very small and borne on thin side-shoots (August/September). Female cones are borne on fairly thick stalks near the branch ends. They are roughly round, up to about 2,5 cm in diameter,

Female cones

dark brown and have conspicuous protuberances. They are persistent, and cones in various stages of development may be found throughout the year.

(Zimbabwe: (10) Mountain cedar)

9

 The two *Hyphaene* species of palm are geographically distinctly separated. *H. coriacea* (previously *H. crinita* and *H. natalensis*) occurs along the coasts of Transkei, Natal and Mozambique, and also inland in Mozambique, the Transvaal Lowveld and eastern Zimbabwe. *H. petersiana*, (previously *H. benguellensis*) inhabits the northern areas of Namibia, Botswana, Zimbabwe, northern Transvaal.

The most conspicuous difference between the two species is the shape of the fruit. In the case of *H. coriacea* it is supposed to be *pear-shaped*, and in the other, round (Palgrave, 1983). In this case, the trees in the Transvaal Lowveld could also be *H. petersiana*.

Fruit

The small reproductive organs are borne in large **inflorescences** with the two sexes on separate trees. The male florets are small, yellow-green and packed in longitudinal or spiral rows. The female florets consist of small, green, spherical knobs, also arranged in rows. Flowering takes place in November/December. The dark brown **fruit** is borne in large pendent clusters. It is relatively large (6 cm in diameter), shiny and very hard when ripe, and reaches maturity after approximately two years. The large **leaves** are fan-shaped, and the petiole has sharp, hooked thorns.

Sap collected from the stems is used to brew an alcoholic beverage. In the process the stems are cut off and eventually die. This is one of the reasons why full-grown trees are somewhat rare.

(Zimbabwe: (16) Southern ilala palm)

 Members of the *Aloe* genus are found mainly in Africa. All of them have characteristic *succulent leaves*, and although quite a number have been incorporated in the national tree lists, only a few are truly tree-like, with a main trunk, conspicuous branches and a crown.

Flowers

On the eastern side of the subcontinent, *A. bainesii* is the outstanding example in this category while on the arid, western side it is the quiver tree. The quiver tree has a relatively restricted distribution in the northern Cape and adjoining Namibia to a point north of Windhoek. It is protected in several national parks in both countries. The common name, quiver tree, derives from the fact that Bushmen hunters fashioned quivers for their arrows from the soft branches. According to Palgrave (1983), Simon van der Stel was the first to record this species, in 1685.

Bark

In full-grown trees *the trunk is massive, measuring about 3 m in circumference at ground level but tapering rapidly further up*. It branches profusely and although the **leaves** are fairly small, situated only at the tips of the branches, the crown is relatively dense. The dark yellow **bark** flakes off in patches, causing the trunk to appear mottled. The branches are grey and smooth. The pale yellow **flowers** are borne on short spikes in branched, axillary inflorescences (June/July). Insects, birds and primates are attracted to the nectar. The **fruit** resembles that of other *Aloe* species, being roughly oval with six distinct, longitudinal grooves. The wood is fibrous and useless.

11

Leaves   *Flower*

 The Natal wild banana has the widest distribution of all the *Strelitzia* species in southern Africa, extending all along the coast from East London to southern Mozambique. It then reappears again in the Chipinga/Mutare region of Zimbabwe and the adjoining area in Mozambique. Two other tree species, *S. alba* and *S. caudata* are limited in their distribution to the forests on the escarpment in Swaziland and the Transvaal and to the coastal region of the southern Cape (Knysna) respectively.

Stem

The three species are very similar and difficult to separate vegetatively. *S. nicolai* may reach about 8 m in height and due to suckering, grows in large clumps. The **leaves** are banana-like with long, fairly stiff petioles and blades of approximately 2 m long x 0,5 m wide.

The **flowers** are borne in bluish, boat-shaped spathes, each consisting of three white sepals standing upright and three petals, *two of which are bright blue and lie close together to form a structure which resembles an arrow-head*, the third being small and frilled. Unlike the two other species, the inflorescences of *S. nicolai* are layered, multiple structures consisting of up to five spathes – each new one arising from the previous one. A slimy mucilage is produced within the spathe. The **fruit** is a three-lobed, woody capsule and each black seed is capped with a bright orange aril.

This is a beautiful garden subject and is widely used in the subtropical regions of the subcontinent.

(Zimbabwe: (34) Wild strelitzia)

 At present 30 species of fig trees are recognized in southern Africa. Some species are very characteristic, but others vary to such an extent that even expert botanists find it difficult to identify them with certainty. This particular species does not pose a problem as its leaves and

Fruit

figs are very characteristic. The **leaves** are simple, ovate-oval, dark green, fairly hard, about 7 x 3 cm, are conduplicate upwards and *distinctly cordate (heart-shaped) at the base*. The **fruit** is a small (6 mm in diameter) fig, *shining red when ripe* and borne singly or in pairs in the leaf axils. Although the figs are edible they are not really eaten by humans but are an important food source for birds, monkeys and baboons. The tiny flowers are inside the fruit.

One well-known example of this species is the enormous *Wonderboom* (wonder tree)

Bark

that grows at the foot of the Magaliesberg mountain in Pretoria and must be hundreds of years old. All around the mother tree, new stems have grown from branches which became rooted while lying on the ground. The original trunk died long ago, but in 1985 the offspring consisted of 14 groups of stems comprising 72 single stems, each in excess of 10 cm in diameter, roughly arranged in two circles. At that time, a third circle had started to develop (Von Breitenbach, 1985). Further statistics on this tree were published in the *Journal of Dendrology* (Vol. 3 x 4, 1984), before one of the groups of stems was blown over: girth 16,70 m; height 22 m; spread 53,3 m; cover 2 233 m^2.

(Zimbabwe: (60) Wonderboom fig)

The genus *ficus* apparently came into existence in the Cretaceous period, more than 100 million years ago (Von Breitenbach, 1982). During this time a fascinating process between the fig species and a series of wasp species developed, resulting in every fig species now having a specific wasp species as its pollinator. This phenomenon led to a recent reclassification of the *Ficus* genus partly based on the wasps involved in pollination.

Leaves

The sycamore fig is a common sight in and along the rivers of northern Natal, Swaziland, the Transvaal Lowveld and northern Transvaal, Mozambique, Zimbabwe, eastern and northern Botswana as well as northern and central Namibia. The yellow **bark** and *very thick trunks* of large specimens are unmistakable and can be recognized from a distance. Narrow buttresses developing at the base of old trunks are also very characteristic. The large **leaves** are slightly rough, with entire margins.

Fruit

Unlike the majority of fig species, the **fruit** of this species is borne in masses on fruiting branchlets on the trunk and main branches. The figs are fairly big (up to 4 cm x 3,5 cm) and yellow to reddish when ripe; they are an important food source for animals favouring riverine vegetation.

The tough, pale brown timber is light and soft, and is used for making drums.

(Zimbabwe: (65) Sycamore fig)

Bark

 As suggested by the common name, this tree grows only in or near water. It is a tropical species extending as far south as southern Mozambique and northern Natal. It also occurs intermittently in localized areas further north in Mozambique, as well as Zimbabwe, northern

Fruit

Botswana (Moremi swamps) and the Caprivi area of Namibia.

The water fig is the equivalent of the chameleon among the tree flora because of its varied growth forms. In gallery forests it may be a fairly large, sparsely branched tree (12 m), while on the islands in the Okavango Delta it is a smallish, multi-stemmed, much-branched tree or shrub forming dense thickets. The **bark** is quite smooth and pale grey.

It seems to be at least semi-deciduous, possibly deciduous. The simple **leaves** are oval to oblong, thick, fairly hard and leathery, glabrous, a glossy green above and paler beneath, and the underside is sometimes speckled with small encrustations. The margin is rolled under. The conspicuous midrib is yellowish although, as with the petiole, often reddish.

The **fruit** is a small fig, measuring approximately 1 cm in diameter. The figs are borne in pairs in the leaf-axils or immediately above leaf-scars lower down on the twigs, and are roughly spherical, glabrous, glossy and a *bright dark red when mature* (January to August). The male and female **flowers** are borne inside the figs, as is the case with all fig species. The figs are tasty and are favoured by birds and animals, as are the leaves which are highly nutritious. The bark is valued for its medicinal properties.

(Zimbabwe: (67) Water fig)

Shortly after his arrival in the Cape (about 1660) Jan van Riebeeck found it necessary to protect the settlers' cattle from theft, so a hedge of wild almond was planted around the main compound on the foothills of Table Mountain. Parts of this hedge can still be seen at Kirstenbosch and on Wynberg Hill pointing to a lifespan of more than 330 years for the species, which is not the case in most other members of the Proteaceae family.

The wild almond is restricted to the winter rainfall area of the Cape: the southern and south-western coastal areas, where it grows in valleys and often along streams. It is a smallish tree (up to 8 m) with wide-spreading branches originating low down on the trunk and usually forming a nearly impenetrable thicket on the ground. The **leaves** are very hard with sharply-tipped teeth on the margins. *The arrangement of the leaves is rather extraordinary since they occur in whorls of four to six and even on fairly thick branches.* The species' name refers to this phenomenon.

The small, white **flowers** are sweetly scented and are borne in long (up to 8 cm),

Bark

Flowers

showy spikes which occur in the axils of the youngest leaves (December/January). I have not been able to photograph the **fruit**. Palgrave describes it as 'almond-shaped, up to 4,5 x 3 cm and densely covered with rusty-brown, velvety hairs'. Young fruits are maroon. The **bark** is smooth, sometimes striated, and varies from yellowish-grey to pale greyish-brown. The wood is fairly hard, light brown and has a reticulated pattern. Due to a lack of fairly large, straight logs it is apparently not put to use.

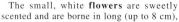

This species occurs from southern Natal, through Swaziland into the Transvaal where it is widespread from the southern Lowveld along the Drakensberg to the northern Transvaal, and to the west as far as southern Botswana. It covers most of the inland areas of Zimbabwe and also crosses into Mozambique and Malawi.

F. saligna occurs most frequently as a smallish tree of 8 m, being very slender when young. A spreading, fairly sparse crown develops with age. Young branchlets are pendent and *conspicuously red*. Old **bark** is grey to nearly black, and deeply fissured longitudinally. The long, narrow **leaves** are drooping, up to 16 cm long, green to yellowish-green and shiny. The petioles and main veins of younger leaves are red. The **flowers** are silver-grey with a red tinge, and are densely packed in cylindrical, pendent spikes of up to 15 cm long (August/January, depending on rainfall). The **fruit** is a small, brown, hairy, nut.

Bark

During the flowering season it produces large quantities of nectar. The wood is pale to dark brown with a red tinge, and is beautifully figured. It is used for cabinet-making.

(Zimbabwe: (78) Beechwood)

Flowers

Flower heads   *Bark*

This is a somewhat rare species which, like the overwhelming majority of the members of the Proteaceae family, is confined to the winter rainfall region of South Africa. This particular subspecies is confined to a relatively small area in the south-western Cape, and the tree illustrated was photographed on the rocky coast near Betty's Bay. The other subspecies, *conocarpodendron*, has an even more limited distribution, being restricted to the Cape Peninsula.

L. conocarpodendron subsp. *viridum* may be classified either as a shrub or as a tree. It can grow to as much as 5 m in height, branches either at ground level or just above and has a roundish, umbrella-shaped and fairly dense crown. The **bark** on the branches and trunk is smooth and greyish-brown, even on older trees. The *branches are characteristically crooked* and young branches are distinctly hairy.

It is evergreen, and the **leaves** are borne close together at the tips of the branches. They are oblong to obovate, up to about 10 x 5 cm and are mostly covered with soft hairs. The basal part of the margin is entire but a number of prominent teeth occur on either side of the apex. The margins of the teeth may be reddish. The beautiful **flower heads** are large, terminal, solitary or in groups of two or three, and consist of a large number of closely packed, bright yellow flowers. Flowering starts in August and may continue in to January. The **fruit** is a very small nut.

Of the 47 species in this genus represented in South Africa, only about seven reach tree size.

This species is the most common Protea in South Africa (Rourke, 1982). Its range extends from the Katberg mountains in the eastern Cape, northwards through Transkei, Lesotho, the Orange Free State (east and north), Natal, Swaziland and into the Transvaal

Flower heads

where it occupies most mountains. It favours well-drained, sandy soils. It does not feature in the Transvaal Lowveld, but an isolated population exists in eastern Zimbabwe, on the border of Mozambique.

It occurs most frequently as a *single-stemmed tree of up to 8 m high*, with a short trunk and a sparse, spreading crown. The fissured **bark** is dark grey to black, and breaks up into small blocks. Young branches are smooth and pale grey. The **leaves** are usually linear-elliptic, up to 25 cm long, glabrous, greyish-green, thick, hard and covered with silvery hairs.

Bark

Flower heads are usually borne solitary and terminally (December/January). They are mainly pink or carmine, with green at the base, although forms with creamy-green bracts are also found. They have a pleasant, sweet odour. The **fruit** is a small nut. Large specimens may be up to 100 years old. It grows slowly, but this tree is highly recommended as a garden subject, and should start flowering at about five years of age.

(Zimbabwe: (83) Manica protea)

Flower head

Bark

Inclusion of at least one of the *Proteas* is an absolute necessity for any publication dealing with the woody vegetation – whether trees or shrubs – of southern Africa. The choice of this species was facilitated by the fact that it is considered to be *'the tallest growing of all the arborescent proteas* in South Africa' (Rourke, 1982). Although it was collected by Burchell in 1814, it was named in honour of Leopold Mund who arrived at the Cape in 1815 to collect plants for the Berlin Museum.

This has a rather curious distribution. It occupies predominantly the area between George and the Groendal Wilderness area in the Strydomsberg, near Uitenhage in the Cape. It is abundant over this range, which includes the Tsitsikamma area. About 300 km to the west, between Hermanus and Betty's Bay, several isolated populations flourish, with no recorded occurrence anywhere between the two areas.

This is usually a smallish, upright, slender tree measuring up to approximately 8 m in height, but it may reach as much as 12 m. The main trunk is short as it subdivides near the ground. The **bark** is smooth and greyish-brown, and the branches are long and upright.

The **leaves** are narrowly-elliptic to elliptic, up to 12 cm long, pale green and distinctly veined, with the midrib often a reddish colour. The **flower heads** are up to 8 cm long, do not open wide, and have silky green bracts fringed with hairs. (The bracts may also be pink.) Flowering commences in midsummer and may continue until the following spring. The **fruit** is a small nut covered in hair-like outgrowths.

Only two tree species in the Sandalwood family (Santalaceae) are present in southern Africa, the other being *Osyris lanceolata* (Transvaal sumach). As the latter is cold-resistant, it is widespread from the eastern Cape, through the Orange Free State and Lesotho, into the Transvaal. It is also found in all the other countries of the subcontinent. Its leaves are markedly blue-green and smallish while the fruit is small and brilliant red with a circular mark at the apex.

Flowers

The Cape sumach is widespread – mainly confined to the coastal dunes, but also occurring fairly far inland – from the western Cape all along the coast to northern Natal, and possibly southern Mozambique as well.

In fairly open fynbos areas on the coastal dunes it is a *very dense, spreading, multi-stemmed, small tree of 3-4 m, with its branches on the ground*.

Fruit

The stems have fairly smooth, brownish-grey **bark**. The **leaves** are smallish (up to 5 cm long), elliptic, leathery and hard, dull green with a grey bloom and sharply tipped. They are borne close together in opposite pairs.

The tiny **flowers** are inconspicuous and yellowish-green, and are borne in compact, terminal heads. Flowering takes place over a very long period, probably March to September, and flowers and mature fruit are usually found together. The oval **fruit** is often closely packed and may be up to 1,5 cm long. It becomes shiny red and changes to purplish-black when mature.

Due to a high tannin content, the leaves and bark have been used for the tanning of hides.

 This is a tropical species with an easterly distribution, occurring along the Zambezi in Mozambique, eastern Zimbabwe, the north-eastern corner of the Transvaal, and with an isolated population at the southern tip of Mozambique and Maputaland in Natal. It grows in deep sand, occasionally on rocky outcrops and usually in dense thickets.

It is a multi-stemmed, deciduous, slender plant with long, thin, supple branches. The **bark** is dark greyish-brown and smooth. The exceptionally glossy **leaves** are bright green, oblanceolate to nearly elliptic, up to 15 cm long and are borne alternately. The showy **flowers** are *beautiful and unusual*, borne solitary near the ends of the branches (September/November). They consist of three sepals and six petals, in two whorls of three. Those at the base (the outer whorl) are broad, flat, spreading and bent either backwards or forwards. Those in the inner whorl stand upright and close together, and are cupped and clawed. Different colour variations occur. The **fruit** is reminiscent of the custard apple, nearly spherical, dark green mottled with white and up to 7 cm in diameter. At maturity it turns brownish-black and wrinkled.

(Zimbabwe: (119) Green-apple)

Fruit

Flowers

 This tree belongs to the Annonaceae family, most members of which are noted for their *extraordinary, very brightly coloured, edible fruit*. It is most often encountered as a shrub or even a scrambler, but may possibly grow to about 7 m. Nearly all plants are multi-stemmed, with the branch-ends on the ground.

Fruit

Although the savanna dwaba-berry grows in fairly divergent habitats, it seems to thrive in fairly dense woodland on well-drained sand, as is evident by the fact that dense thickets often develop. In order to see this species, one must either visit the far northern part of Botswana, the Caprivi Strip or the northern, eastern or north-western areas of Zimbabwe. It is also known to occur in the north-western part of Mozambique.

The **bark** is grey and fairly smooth. The **leaves** are simple, markedly obovate to nearly rectangular, thin, fairly large (mostly about 8 x 5 cm), and are always velvety when young, while only occasionally velvety when mature.

The inconspicuous, cream-coloured **flowers** are borne solitarily and have thick, rather fleshy sepals and petals (November/December). Its characteristic feature, the red, fleshy, edible **fruit**, is very attractive. It occurs in bunches reminiscent of a hand, or of miniature sausages strung together. The fruit is cylindrical, measures up to about 8 cm long and is constricted between the seeds. Primates, birds and insects relish it.

(Zimbabwe: (110) Northern dwaba-berry)

 This tree occurs in the north of Namibia and Mozambique, the north and south of Zimbabwe, and in South Africa at Crook's Corner in the north-eastern Transvaal. It prefers a rocky substrate and is found on hillsides or ridges.

The propeller tree always has a single, tall, bare, exceptionally smooth and shiny trunk. On the shady side the **bark** is usually pale grey to grey-brown, but on the sunny side of the trunk it is generally bleached almost white. It is a smallish tree (8 m), but may reach 15 m and has a sparse, spreading, more-or-less rounded crown.

Fruit

The simple **leaves** measure 10 x 10 cm; older leaves have three lobes. The small, yellow **flowers** are borne in terminal racemes, each containing several male and fewer female flowers in autumn. The **fruit** is borne in pendent clusters and changes colour from light green through yellow to dark brown. Each consists of an oval, very hard fruit about 2 cm in length, with *two stiff, narrow, papery wings* of about 7 cm in length. When the ripe fruit falls in July/August, the weight of the nut turns it around and the wings cause it to spin as it floats down.

(Zimbabwe: (122) Propeller tree)

Flowers

 Although it has a widespread distribution, this tree seems to prefer arid conditions. It occurs throughout Namibia, the northern Cape, the western Orange Free State, the entire Transvaal except the eastern Highveld, Swaziland, some areas of Natal, southern Mozambique and the southern and western parts of western Zimbabwe. It occurs on a variety of soil types, from sand to clay and even rocky locations in open or fairly dense woodland.

Fruit

While it may reach a height of 10 m in an ideal situation, it is usually only between 5 and 7 m high. It is a single-stemmed tree with a small, rather dense, twiggy crown. The branches are very smooth and almost white. The stem is smooth, occasionally grooved, and the **bark** may be smooth and almost white or greyish-brown, and peeling in small sections. It is evergreen to deciduous.

Flowers

The **leaves** are small, simple and borne singly or in small groups on abbreviated lateral twigs. They are thick, hard, brittle and slightly scabrid. The **flowers** are greenish-yellow, minute and are borne in racemes in such great quantities along the twigs and thicker branches, that *the tree is quite imposing in full bloom* (usually October/December, but sometimes later). If conditions are right, large quantities of **fruit** are produced. The fruit is spherical, smooth and yellow to pale red when ripe. It is edible and is relished by birds.

(Zimbabwe: (130) Shepherd's tree)

 Despite its wide distribution in the subcontinent, the bead-bean is nowhere really abundant. In South Africa individual specimens usually occur very far apart, and it is therefore one of the lesser-known components of the tree flora. It may reach about 15 m in height but is usually not more than about 6 m. It has a single, bare trunk and a moderately spreading, sparse crown. One leafless tree has been found, but the bead-bean is probably evergreen. Old stems are grey-brown with a red-brown tinge. The **bark** has shallow grooves and peels off in small, powdery flakes exposing the bright green, living bark.

Fruit

The simple **leaves** are borne spirally at the twig terminals. The most characteristic part of the leaf is the petiole, which is reddish-brown, bent at a distinct angle near its conjunction with the leaf and cylindrically thickened at both ends. Like most members of the Capparaceae family, petals are absent, the most obvious component of the attractive **flowers** being the *long, brilliantly white stamens* which turn yellow when withering (September/October). They are borne solitary and axillary. The **fruit** is cylindrical and conspicuously deeply-segmented, resembling a pod.

Flowers

(Zimbabwe: (142) Bead-bean)

 One of the major wild fruit trees in southern Africa, the mobola plum is left untouched by the African people when bush is cleared for crops. It is widespread almost throughout Zimbabwe. It is also found in the Transvaal Lowveld, Swaziland, the Caprivi Strip and northern Mozambique.

Reputed to reach a height of 24 m, most information indicates an average height of roughly half that size. It is a single-stemmed tree with a short, bare trunk and a dense, wide-spreading crown characterized by drooping branch ends. It is practically ever-green. The **bark** is rough and dark grey, although in areas where veldfires occur frequently, the bark is always black. It grows on deep, well-drained sandy soils only.

Bark

The simple **leaves** measure 9 x 5 cm and are hard and brittle. Older leaves are dark green, glossy and glabrous on the upperside but the undersides are covered by an off-white or rust-coloured downy layer. The secondary veins are arranged in a *herring-bone pattern*. The small, white **flowers** are borne in large, axillary or terminal panicles (October/ November). The oval **fruit** (up to 3 cm long) is edible and sweetly flavoured.

Fruit

The pulp is yellowish when the fruit is ripe (August/September).

(Zimbabwe: (166) Mobola plum)

27

Bark   *Fruit*   *Flowers*

This tropical tree species is widespread in central Africa, but on the subcontinent occurs only in the east: Mozambique, the eastern and south-eastern parts of Zimbabwe, the Transvaal Lowveld, and along the coast of Natal and Transkei. It is prominent along the coast.

The common name is very descriptive as most of the full-grown trees have the same unmistakable, *wide-spreading and mostly flat crown supported by a long, straight, bare trunk*. It is deciduous, but is densely covered with leaves during summer. The biggest trees I have encountered are about 18 m high, but Palgrave (1983) mentions a maximum height of 40 m.

The **leaves** are bipinnately compound and up to 25 cm long. The **flowers** emerge in spring, after the new leaves, and are borne in dense heads at the tips of the twigs. They are less attractive than those of most other *Albizia* species. The **fruit** is a fairly long (15 cm), narrow (2,5 cm), flat pod. It is biscuit-coloured to pale brown when mature, dehiscent and only reaching maturity in the autumn of the following year.

Although the wood is relatively soft and light, it is suitable for the manufacturing of furniture as well as for parquet floors.

(Zimbabwe: (172) Rough-bark flat-crown)

This is a tropical species occurring throughout Africa from Swaziland to Tanzania, including the Transvaal Lowveld, northern Transvaal, eastern and northern Botswana, north-eastern Namibia and various parts of Zimbabwe and Mozambique. It grows on a variety of soil types, but more and larger trees are found on low-lying alluvial soils and brackish plains.

Although often encountered as a small, very sparse, multi-stemmed shrub, this tree can reach a height of 16 m, as shown in the picture. The girth at breast height of that particular tree's trunk is 4,10 m. The trunk usually branches fairly low and the somewhat dense crown is widespreading. It is deciduous. Old stems are dark grey and the rough **bark** cracks into prominent, vertical ridges which fuse at random intervals.

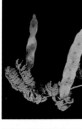

Fruit

The **leaves** are feathery, bipinnately compound and fairly long and narrow (15 x 5 cm). The leaflets are grey-green, small (6 x 2 mm) and slightly sickle-shaped. The small **flowers** are white and are borne in rather loose, fluffy heads, with or just after the young leaves (October/November). As with other *Albizia* species, the stamens are the most important and visible component of the flowers. The **fruit** is a *pale brown, pergamentaceous, thin, flat and oblong pod up to 13 cm long and 3 cm wide.* Usually masses of pendent pods, singly or in small bunches, are produced.

The timber is medium-heavy, pale brown and finely textured and produces a smooth finish but is not used for any commercial purpose on a large scale.

(Zimbabwe: (180) Sickle-leaved albizia)

29

Albizia versicolor (SA 158) Large-leaved false-thorn

This tropical species occurs roughly from northern Namibia eastwards into the Caprivi Strip and northern Botswana, and then along the Zambezi to the Indian Ocean. From north-eastern Zimbabwe and northern Mozambique it extends southwards through the Lowveld into Swaziland and northern Natal. It shows a preference for moist conditions and is often found in low-lying areas near watercourses. In areas with a relatively high rainfall and/or deep, sandy soil, it also occurs in fairly open woodland. It is a medium to large deciduous tree (up to 18 m) with a single, straight, long trunk and a dense, wide-spreading crown. *Young twigs are covered with golden-brown hairs.* Old stems are rough and dark grey and the **bark** peels off in small, flat sections. The **leaves** are large (about 30 x 20 cm) and bipinnately compound. New leaflets are dark reddish brown and old ones green to slightly dark green. The leaflets are big (up to 5,5 x 3,5 cm) and vary from broadly elliptic to almost rectangular. The powderpuff-like **inflorescences** are large and white but wither quickly and become yellow (November/ December or later). The **fruit** is a thin, flat, oblong (up to 20 x 5 cm) glabrous and smooth pod, changing in colour from green to yellowy-green, to wine-red and finally pale brown.
(Zimbabwe: (185) Poison-pod albizia)

Inflorescences

Bark

Faidherbia albida (SA 159) Ana tree

The public was introduced to the existence and size of this tree by the famous South African author Eugene Marais, who wrote about the gigantic trees on the bank of the Magalakwin River in the north-western Transvaal.

The ana tree occurs in the warmer, mostly frost-free areas of South Africa, from northern Natal through the Transvaal Lowveld to the northern parts of the Transvaal. It is relatively rare in Botswana but is fairly abundant in areas such as the north-western parts of Namibia and the extreme northern region of Zimbabwe. Although accustomed to a mild climate, it can withstand very low temperatures. It grows mostly on the banks of rivers and other watercourses.

This species has recently been removed from the *Acacia* genus and is now the only species in the newly-created genus *Faecherbia*. One aspect in which it differs markedly from the *Acacias*, is the fact that it sheds its *leaves in summer*. New leaves emerge immediately afterwards, with the result that the tree is in full leaf in winter. The **leaves** are bipinnately compound and bluish-green.

The white **flowers**, which are borne in long spikes, also emerge in autumn, which is the 'wrong' time for most African trees. The **fruit**, a large, attractively coloured pod, is well known for its nutritional value.

(Zimbabwe: (189) Apple-ring thorn-tree)

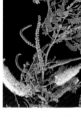

Flowers

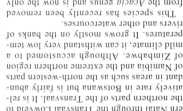

Fruit

Acacia erioloba (SA 168) Camel thorn

The camel thorn is undoubtedly the king of trees in the arid western regions of southern Africa, although it also occurs in the central area of the Transvaal, the south-western part of Zimbabwe and the north-ern area of Botswana where the rainfall is relatively high. In these areas the tree has a widespread distrib-ution in a variety of ecological situations, but it tends to prefer sandy soil. In the north-western Cape and western Transvaal it is often the dominant tree on the plains, sometimes occurring in dense, isolated clumps. In very arid areas it grows almost exclusively in the beds or on the banks of rivers. Trees occurring outside this favoured habitat are usually stunted.

Although gigantic trees (height 18 m, girth 9 m) have been reported from the Kuiseb River in Namibia (Theron, 1984), they gen-erally reach between 9 and 10 m in height. The **bark** is rough and longitudinally fissured. The shiny, initially *dark brown and later grey, swollen, straight thorns*, which occur in pairs, are characteristic. So too are the bright yellow, spherical **inflorescences** (August/September). The **fruit** is a pod, usually very broad, thick, sickle-shaped, grey and velvety. It is indehiscent. Variations occur in the shape and size of the pods, with some being thin, round and long. The small **leaves** are bipinnately compound.

This is a valuable evergreen tree, especially in the hot, dry, west-ern areas, as it provides shade throughout the year. The leaves and pods, which are rich in protein, sustain a variety of wild animals and domestic stock.

(Zimbabwe: (198) Camel thorn)

Inflorescences

Fruit

Acacia hebeclada subsp. hebeclada (SA 170) Candle thorn

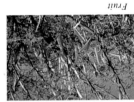

Fruit

Inflorescences

This tree is extraordinary for two reasons. The first involves the setting of the pods, *which stand upright,* unlike all other pods, which are pendent. Even more intriguing is that the same applies to the second sub-species, *chobiensis,* but not to the third (*trisits*).

The second phenomenon involves the growth habit of subsp. *hebeclada,* which is the major subspecies. Although it is a true tree-like plant, branching low in some cases, it occurs more often as a dense thicket which may assume very large proportions as new plants sprout from the roots around the mother plant. The largest of these, included in the *National Register of Big Trees,* has the following dimensions: girth of all stems in total 6.06 m; height 7 m; spread 55.6 m and cover 2 431 m². This particular tree grows in the Waterberg area of the Transvaal, which has a relatively high rainfall. In the main distribution area of the tree (the arid western regions of the subcontinent, including the northern Cape, northern, eastern and southern Botswana and large parts of Namibia) it can assume the same spread and cover as the specimen above, but will be only 1 or 2 m high.

The **bark** is rough and longitudinally fissured, and the **leaves** are bipinnately compound. The small, off-white, spherical **inflores-cences** are produced from July to September. The **fruit** is a long, straight, thick, hard, brown pod covered with yellowish hairs. Bushmen build their shelters in the middle of these thickets to protect themselves from lions and other predators.

(Zimbabwe: (206) Candle-pod acacia)

Acacia karroo (SA 172) Sweet thorn

This is one of the most common trees in the greater part of the subcontinent. It has therefore adapted to large fluctuations in temperature and moisture. With the exception of the Gordonia area in the northern Cape and the extreme north-eastern corner of the Transvaal, it can be found throughout South Africa. It also covers the greater parts of Namibia and Zimbabwe. In Mozambique it is limited to the eastern, south-eastern and in Botswana to the eastern, south-eastern and northern regions.

Flowers

Sweet thorns occur on a variety of soil types and in a variety of habitats. In a number of areas, such as the Karoo, Namibia and north-western Transvaal, it is nevertheless most often encountered in the low-lying areas with clayey soil, next to watercourses.

In growth form, this tree varies considerably. In the Karoo, Orange Free State and western Transvaal it is generally a smallish tree with a short trunk and a spreading crown. In other areas, such as the north-western Transvaal and the Lowveld, it is a long, slender tree. In all areas it often occurs in *dense, nearly homogeneous stands*. The **bark** is smooth and brownish-grey, and on old specimens becomes black and longitudinally fissured. The dark-green **leaves** are bipinnately compound. The beautiful, dark yellow, spherical **flower heads** appear mostly in November/December but sometimes as late as March. The **fruit** is a sickle-shaped, dehiscent pod, and remains on the tree for a year or more.
(Zimbabwe: (208) Sweet thorn)

Acacia tortilis subsp. *heteracantha* (SA 188) Umbrella thorn

Two distinct sub-species of this tree occur in southern Africa. Subspecies *heteracantha* is widespread in all the countries of the subcontinent and it is adapted to a wide range of climatic conditions.

It is deciduous and usually not more than 10 m high. In north-ern Botswana (Moremi) some trees reach 20 m. The well-known, conspicuous umbrella-shaped crown only develops fully in old specimens, young trees having roundish or flat-topped crowns. The stem is usually fairly short and the main branches bare. Old **bark** is dark grey to black and fairly deeply, longitudinally fissured and ridged. *The very sharp spines* of this plant are unique since some are short, blackish and hooked, while others are long, white and straight. (The species name refers to the spines.) The spines occur in pairs, mostly two of the same kind together but sometimes mixed (thus a hooked and a straight spine together). The bipinnately compound **leaves** are probably the smallest among the thorn trees (3 x 1,5 cm). The same applies to the leaflets, which are minute (1,5 x 0,5 mm). The inflorescences are globose and consist of a fairly large number of small, white **flowers** which are packed together. They are borne in groups among the leaves (usually November/December). The **fruit** consists of characteristic, pale brown pods, which are always spirally contorted and some-times intertwined with each other. All parts are without hairs.

The heartwood of umbrella thorn is reddish-brown, very hard and fairly heavy. It is seldom used except as firewood. The leaves and highly nutritious pods are widely foraged by a variety of wild animals as well as by domestic stock. The pods are mainly uti-lized during winter, after they have fallen to the ground.

(Zimbabwe: (226) Umbrella thorn)

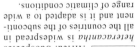

Fruit

Acacia xanthophloea (SA 189) Fever tree

Inflorescences

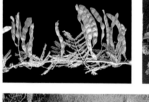

Fruit

Fever trees are known worldwide, possibly as a result of Rudyard Kipling's story about the elephant's child, who continued travelling northwards until at last he came 'to the banks of the great, grey-green, greasy Limpopo River', all set about with fever trees'.

It has a markedly easterly distribution in Africa, from northern Natal through Swaziland, Mozambique, the eastern part of the Transvaal Lowveld, the lowveld of Zimbabwe and further north. They grow in low-lying, fairly wet to swampy areas, and may occur in very dense stands.

It is an unusual, striking, medium-large, deciduous tree (up to 15 m) with a long, single stem and wide-spreading crown. The **bark** of old stems is very smooth, yellow-green and powdery, with conspicuous longitudinal indentations. Its long, white *thorns are straight, smooth, sharp and are borne in pairs*. The bipinnately compound **leaves** are rather small, and occur in small groups immediately above the thorns. The **inflorescences** are fragrant, golden-yellow, round, and are borne among the leaves (August/September). The **fruit** is a pod of up to 10 cm long.
The timber is pale brown with a reddish tinge and is quite hard and heavy. It can be used for carpentry.
(Zimbabwe: (228) Fever tree)

Neltonia hildebrandtii var. *hildebrandtii*

(SA 191) Lebombo wattle

In South Africa this species occurs mainly in Natal. Literally dozens of beautiful specimens grow in the sandveld next to the road to Kosi Bay, east of the Pongola River. The only other area where it occurs is in the Lebombo mountain range, immediately north of the Olifants River in the Kruger National Park, an area that is inaccessible to the general public. It also features fairly widely in Mozambique, as well as south-eastern Zimbabwe and isolated patches further north.

Flowers

It is a large (up to 25 m), deciduous tree with a widespreading, dense crown and a relatively short, bare trunk. The branch-ends are pendent. Old stems are dark grey, and the **bark** peels off in strips and small blocks. The **leaves** are bipinnately compound, fairly small (9 cm long) and dark green. A small, brown gland occurs on the rachis between each pair of pinnae. The small, white **flowers** are grouped together in fairly long (8 cm),

Bark

axillary spikes. They are produced in large quantities (November/December, depending on rainfall). The **fruit** is an exceptional, *flat and surrounded by a reddish-brown papery wing*. The pods split on one side only to release the seeds, but the seeds remain attached to the pod for months on end.

The timber is hard, heavy and dark brown to black. It is of good quality, and should be suitable for cabinet-making.

(Zimbabwe: (233) Lowveld newtonia)

Burkea africana (SA 197) Red seringa

Bark

Fruit

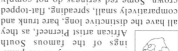

My guess is that this is the tree featured in many of the paintings of the famous South African artist Pierneef, as they all have the distinctive long, bare trunk and comparatively small, spreading, flat-topped crown. Some red seringas do not comply with this description, such as the example in the Potgietersrus district, which has a fairly short trunk and a wide-spreading crown (23,5 m diameter), as measured by the Dendrological Society.

The red seringa occurs in the Transvaal, from Johannesburg northwards throughout Zimbabwe, large areas of Mozambique, the northern and southeastern regions of Botswana and northern Namibia.

It is a deciduous tree, which are close together at the twig terminals, confined to the tree-tops. The young twigs are *distinctly rust-coloured* due to a dense cover of hairs, and the old **bark** is dark grey and subdivided into small blocks. The **leaves** are bipinnately compound with two (sometimes three) pairs of pinnae. The small **flowers** are sweet-smelling, white to creamy and are grouped together in single, long (25 cm) spikes borne in the leaf axils (October/November). The **fruit** consists of large quantities of pods in pendent clusters. They are smallish, flat, hard, woody and single-seeded.

The durable timber is hard, tough and pale red to red-brown.

(Zimbabwe: (243) Burkea)

38

Flowers

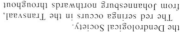

Colophospermum mopane (SA 198) Mopane

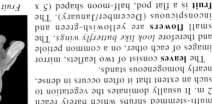

The mopane dominates the African vegetation over large tracts of land. In southern Africa this applies to the middle section of northern Namibia, the Caprivi Strip, north-eastern Botswana, the west, south and north of Zimbabwe, large areas in central Mozambique as well as the northern and north-eastern parts of the Transvaal – especially in the Kruger National Park and north of the Soutpansberg.

Under favourable soil and climatic conditions, as in Moremi National Park in Botswana, it grows into a large tree (20 m) with a thickest stem and long, bare branches. In contrast, the plants in the north-eastern section of the Kruger National Park are multi-stemmed shrubs which barely reach 2 m. It usually dominates the vegetation to such an extent that it often occurs in dense, nearly homogeneous stands.

The **leaves** consist of two leaflets, mirror images of each other, on a common petiole and therefore look *like butterfly wings*. The small **flowers** are yellowish-green and inconspicuous (December/January). The **fruit** is a flat pod, half-moon shaped (5 x 2,5 cm) wrinkled and biscuit-coloured when ripe. The seeds have small, reddish glands, exuding a sticky fluid on both surfaces.

The wood is hard, heavy and mainly dark brown to nearly black. (Zimbabwe) (246) Mopane.

Bark

Fruit

Guibourtia coleosperma (SA 199)   Large false mopane

This genus is represented by two tree species in the sub-continent of southern Africa. The smaller one (*G. conjugata*) is widespread and deciduous. *G. coleosperma* occurs only in the north-eastern area of Namibia, the extreme north of Botswana and then spreads south-

Botswana and then spreads southeastwards along the border between Zimbabwe and Botswana.

It is apparently evergreen, and is an impressive, single-stemmed tree with a wide-spreading crown. It can grow up to 20 m high. Young branches are reddish, becoming conspicuously cream with dark brown to black patches of flaking **bark**. The bark of old trunks is slightly yellowish-grey or can become dark blackish-brown.

The common name refers to the fact that the identical, sickle-shaped pair of **leaflets** resemble that of the mopane. The smallish, white, star-shaped **flowers** are borne in fairly large, terminal panicles (December/April). The **fruit** is very characteristic. It is a *small, almost circular (up to 3 cm long), flat but thickened pod* which turns brown when mature. When ripe, it splits on one side and the two valves curl back to release the single, shiny, red seed which has a bright yellow aril on one side and is attached to a thread-like stalk. The seeds and arils are widely used as food. The wood is commercially utilized and is known as mochibi. It is attractive, rather soft and pinkish-brown.

(Zimbabwe: (244) Large false mopane)

Fruit

Flowers

Bark

40

All three of the eight *Brachyste-gia* species in southern Africa occur in Zimbabwe and Mozambique. Of these, only *B. boehnii* succeeded in establishing itself in Botswana (in the Kasane area).

With one exception, all the *Brachystegia* species have the true tree form: a single, long trunk and a wide-spreading crown. Because of its local abundance, beautiful shape and the deep red colouring of the young leaves, the msasa (*B. spiciformis*) is the best known among them. Together with *Julbernardia globiflora* and *B. glaucescens*, the two *Brachystegia* species already mentioned largely dominate the highveld vegetation in central Zimbabwe.

Leaf bud

Prince-of-Wales' feathers is a very attractive, large, deciduous tree, which can measure in excess of 15 m. It earned its common name from the striking colours of *the large leaf buds and leaves which are often bright red*. Old **leaves** are feathery and greyish-green. The somewhat inconspicuous **flowers** are greenish-white, sweetly-scented and are borne in short axillary or terminal spikes (September/December). The **fruit**, a panga-shaped pod, is hard, woody and covered with dark brown hairs, especially when young. It splits open explosively while still on the tree to scatter the seeds.

Brachystegia boehnii (Zim. 248) Prince-of-Wales' feathers

Schotia afra var. *afra* (SA 201) Karoo boer-bean

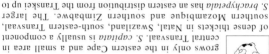

Bark	Flowers	Fruit

Four species belonging to the genus *Schotia*, one with two subspecies, occur in southern Africa. *S. latifolia* grows only in the eastern Cape and a small area in central Transvaal. *S. capitata* is usually a component of dense thickets in Natal, Swaziland, south-eastern Transvaal, southern Mozambique and southern Zimbabwe. The larger *S. brachypetala* has an eastern distribution from the Transkei up to the north of Zimbabwe. *S. afra* var. *afra* is very common in the arid regions of the eastern and south-eastern Cape.

This boer-bean is a small (up to 5 m), particularly dense evergreen tree, often with a *twisted trunk* and smooth, pale grey branches and twigs which are very stiff. Old trunks have dark grey **bark** which remains fairly smooth. The **leaves** are imparipinnately compound and feathery. The leaflets are small, linear and dark green. The striking **flowers** are borne in dense clusters. The petals are rarely pink, and normal flowers are bright red and have a waxy appearance. Flowering takes place in spring (August/October), depending on the rainfall. The **fruit** comprises bunches of large, flat, twisted, colourful pods. Each pod is encircled by a distinctive rim which remains hanging on the tree when the two flat sides break loose at ripening. They become woody and brown when mature. The pale brown seeds are slightly flattened and roundish and *the aril is very small or absent*.

42

Baikiaea plurijuga (SA 206) Zambezi teak

Currently available distribution data locate this tree in the northern and north-eastern regions of Namibia, northern Botswana and the west-ern, south-western and central areas of Zimbabwe.

It has a single, straight trunk and a dense, wide-spreading crown which is mostly roundish at the top, with the outer branch-es often hanging rather low. It prefers the deep Kalahari sand which is the dominant soil type in its distribution areas, and is found mainly in fairly open woodland. Old specimens may be up to 15 m in height. The **bark** is smooth and pale grey on young branches, becoming grey-brown and vertically fissured in old specimens. The **leaves** are pinnately compound with four or five pairs of opposite leaflets.

Inflorescence

Fruit

An outstanding feature of the Zambezi teak is the fact that the *large inflorescences (up to 30 cm long) stand upright,* and can therefore be clearly seen above the canopy. Each **inflorescence** (a raceme) consists of a large number of flowers arranged in two rows on either side of the peduncle. They are fairly large, with four strik-ing purple, and one pale purple to nearly white, petals. The flowers at the base of the inflorescence open first and as only one or two flowers are open simultaneously, the flowering period is extended (usually December/March). The **fruit** is a thick, hairy, woody pod which splits open while on the trees, to release the seeds.

The high-quality wood is reddish-brown, finely textured and durable. It is used for furniture, flooring and even railway sleepers.

(Zimbabwe: (257) Rhodesian teak)

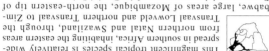

Bark | Leaves | Fruit

This magnificent tropical species is relatively widespread in southern Africa, inhabiting the eastern areas from northern Natal and Swaziland, through the Transvaal Lowveld and northern Transvaal to Zimbabwe, large areas of Mozambique, the north-eastern tip of Botswana and the eastern part of the Caprivi Strip.

It is a single-stemmed, deciduous tree with an *unusually wide-spreading, sometimes umbrella-like crown*. Although it does not often exceed 12 m in height, Palgrave (1983) states that in ideal conditions it can reach 35 m.

The **leaves** are pinnately compound and up to 40 cm long (mostly 15 cm). The **flowers** are extraordinary as they have one petal only. They are nevertheless quite large, and are red with yellow veining or red speckles. Flowering can commence in August but is usually in October/November. The **fruit** comprises large pods (20 x 7 cm) which are half-moon shaped, flat, thick, woody and dark brown. They split on the tree to release seeds which are oblong, shiny and black, with a scarlet to orange aril enveloping one end. They are often strung into necklaces or other trinkets and sold as curios. The wood is well-known in the furniture trade and is sold under the trade name Chamfuti.

(Zimbabwe: (259) Pod mahogany)

Julbernardia globiflora (Zim. 260) Munondo

This is the only *Julbernardia* species occurring in Zimbabwe and Botswana as well as in Mozambique. The other southern African species, *J. paniculata*, has a limited distribution in Mozambique, occurring north of Beira only. Six other species occur in tropical Africa north of the Zambezi. This particular species can be found just about throughout Zimbabwe, but does not enter South Africa, but apparently has an eastern distribution in Africa.

It is a magnificent, single-stemmed, deciduous tree of up to 16 m in height and, according to Palgrave (1983), is an ecologically important species. It is co-dominant with *Brachystegia spiciformis* over large areas of Zimbabwe, Mozambique and countries such as Tanzania and Zaire further north. It has a limited distribution in the north-east of Botswana and the Caprivi Strip.

The **flowers** are white, fairly small and are borne in large (up to 30 cm long), loose, terminal heads. They appear from January to May and are dropped soon after opening. The **pods** are *large, dark brown, velvety and crown the tree. They open explosively while on the tree to release the seeds.* Young **leaves** are soft pink or fawn, as opposed to those of *Brachystegia spiciformis* with which it may be confused, whose young leaves are pink to intensely red. In the case of the latter the pods are also without hairs and are borne among the leaves.

The wood is hard, coarse and very durable. Palgrave (1983) states that it is difficult to work, tending to tear when sawn and split badly when nailed, but it is used quite widely as a general purpose timber.

Flowers

Bark

Bark (young)

Bark (old)

This is not a particu-
larly impressive tree
as it reaches only
5 m in height, but
several of its other characteristics
justify closer attention by gar-
deners. Its distribution in the arid
north-western area of the Cape as
well as central Namibia as far as
the Kunene River, proves that it
is both drought-resistant as well as frost-tolerant. It can therefore
be cultivated in most areas of the subcontinent.

Fruit and flowers

It is a multi-stemmed plant with smooth greyish **bark** on young
branches, becoming dark grey with longitudinal ridges when old.
The crown is sparse, only slightly spreading and with long, slender, *trailing branchlets*. The **leaves** are simple, bluish-green,
smallish and two-lobed like those of the plants in the *Bauhinia*
genus, to which it belonged until recently. The **flowers** are not
spectacular but are beautifully coloured. They are tubular, up to
2.5 cm in length, and the petals are greyish with maroon net-veining and the calyx maroon-red. The stamens are conspicuous. The
pods are also a beautiful reddish colour when young. They are
half-moon shaped, flat and dotted with glands. A flowering time of
September to January has been reported, but all the evidence
points to a longer flowering season, possibly from the end of winter through to the following autumn. The photograph of young
pods was taken in Augrabies Falls National Park early in August.
The wood is used only as firewood.

46

Bauhinia petersiana subsp. *macrantha*
(SA 208,3) White bauhinia

This species has a somewhat more westerly distribution in the subcontinent than most South African trees. From the northern Cape, its range extends northwards through the western side of Botswana into Zimbabwe, where it is widespread. It also extends westwards through northern Botswana and the Caprivi Strip into northern Namibia. In the deep sandy soil of the Chobe/Kasane area of Botswana it is abundant. Further north it can be found in all the countries up to Tanzania and over the entire width of the continent.

When growing in close proximity to other trees, this evergreen plant is a robust climber which can reach as much as 12 m in height. In open woodland it can be a small (up to 6 m), bushy, single-stemmed tree with branches down to the ground. The *butterfly-winged* **leaves** are typical of the *Bauhinias*, being simple and deeply two-lobed. It is interesting to note that the genus was named after the two brothers Bauhin, both botanists, who lived in the sixteenth century. Five *Bauhinia* species have tree status in the region. As can be seen in the photograph, the **flowers** are beautiful with their five long (up to 8 cm), crinkly petals. They appear from December to April. The **fruit** is a woody, sharply tipped pod. It is thickened on one side and can be up to 30 cm long. When mature, the pod splits lengthwise and the two sides spiral.

(Zimbabwe: (263) White bauhinia)

Bark

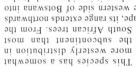

Flowers

Piliostigma thonningii (SA 209) Camel's foot

The camel's foot covers a large segment of the northern part of the subcontinent, from Swaziland to northern Transvaal as well as Mozambique, Zimbabwe, eastern and northern Botswana and the northern extremity of Namibia.

It is mostly a smallish (up to 6 m but up to 10 m in the Caprivi Strip), but fairly conspicuous, deciduous tree with a short trunk and a dense, wide-spreading, roundish crown and pendent branches. The **bark** is pale grey to nearly black, and is subdivided into narrow, but not deeply grooved, longitudinal ridges. The **leaves** are large (up to 12 cm), dark green, glabrous and two-lobed. The trumpet-shaped **flowers** are fairly small and attractive. The crinkly petals are white, sometimes tinged with red at the tips, and the sepals are dark brown due to a dense layer of hairs. Male and female flowers are borne on separate trees (December/February), and look alike. The **fruit** is unmistakable *being a pod of up to 22 x 7 cm in size, 1,5 cm thick, and undulating*. Mature pods are dark brown, woody and indehiscent, but split on the ground.

Flowers

The soft wood is pale to dark brown. It is not used commercially. Extracts of the pods, seeds and roots are used as dye, and the colours vary from red-brown to dark blue or even black.

Fruit

(Zimbabwe: (265) Monkeybread)

Cassia abbreviata subsp. *beareana* (SA 212) Slambok pod

North of southern Africa, this tropical tree species occurs across the entire width of the continent. In the south it is found from north-eastern Botswana through Zimbabwe to the eastern part of the Transvaal, southern Mozambique, northern Natal, Swaziland and in limited areas in northern Namibia.

It is a smallish (10 m) deciduous tree with a single stem and fairly spreading, sometimes roundish, crown. Old **bark** is dark grey to black and rough. The **leaves** are pinnately compound, with up to 11 pairs of leaflets which are thin, dull green in colour and marginally entire.

The **Flowers** are fairly large, measuring 3 cm in diameter. They are dark yellow and are borne in masses at the twig terminals, appearing with or just prior to the new leaves in early spring (August/September). The flowering period is quite short (three to four weeks), which is unfortunate as trees in full bloom are a spectacular sight. The **fruit** is also exceptional; it is a thin pod *which is only 3 cm in diameter, but may be up to about 80 cm long*, and remains on the tree for almost a full year. The fruit pulp is dark green and sticky.

The timber is fairly heavy, hard and dark brown. In frost-free areas, the slambok pod could be cultivated far more frequently. (Zimbabwe: 267) Long-tail cassia)

Flowers

Leaves

Bark

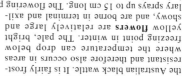

When not in flower this tree is inconspicuous, as it is one of the smaller members of the African flora and does not possess an outstanding or characteristic configuration. Growing to 10 m, it usually has a short stem branching low, with the ends of the lower branches near ground level. It has a very wide distribution, occurring over most of the Transvaal (except the eastern Highveld), northern Natal and Swaziland, southern and northern Botswana, throughout Zimbabwe, most of Mozambique and also in northern Namibia.

Flowers

The **leaves** are bipinnately compound, *feathery, very soft, hairy* and reminiscent of the Australian black wattle. It is fairly frost-resistant and therefore also occurs in areas where the temperature can drop below freezing point in winter. The pale, bright yellow **flowers** are relatively large and showy, and are borne in terminal and axillary sprays up to 15 cm long. The flowering period is exceptionally long (October/February). The petals are crinkled and the outer sides of the sepals are covered with brown, velvety hairs. The **fruit** is a flat, *single-seeded pod with thin, wing-like outer margins*, and it is borne in dense, hanging clusters.

Fruit

The heavy, dark brown wood is used to manufacture various commodities. This is one of the most decorative indigenous trees and should feature more extensively in South African gardens.

(Zimbabwe: (280) Weeping wattle)

Cordyla africana (SA 216) Wild mango

The wild mango is found on the eastern side of Africa from Senegal and the Sudan, to northern Natal. According to available information, it inhabits Swaziland, Mozambique, the Transvaal Lowveld and south-eastern and northern Zimbabwe.

It is a large (up to 25 m), deciduous tree with a fairly long, bare stem and a rather *dense wide-spreading crown*. One specimen on the Komati River has a spread of about 35 m. Very old **bark** is dark grey and rough and peels off in thin, irregular sections. The living underbark is dark green.

The imparipinnately compound **leaves** are arranged alternately on the young branches and are borne horizontally placed on either side of the twig, so that the whole resembles a large, bipinnate leaf. The leaflets are thin, up to 4,5 cm long, light to dark green and shiny. The golden-yellow **flowers** are borne in dense clusters on the basal portion of new twigs just prior to the appearance of the new leaves (September/October). Except in colour, they are almost replicas of those of *Schotia brachypetala*, with the stamens the most important component.

Flowers

The **fruit** (a pod) is drupaceous, glossy, oval to nearly spherical, up to 8 cm long and golden-yellow when ripe. It drops while still green and ripens on the ground. It contains one or two large seeds embedded in a jelly-like pulp which is edible and tasty.

Fruit

(Zimbabwe) Wild mango (281)

Virgilia oroboides subsp. *oroboides* (SA 221) Blossom tree

Although the official common name is blossom tree, most English-speaking South Africans probably know it by its Afrikaans name, *keurboom*. It is endemic to the south-western and south-eastern coastal areas – roughly Cape Town to Port Elizabeth – but it is also fairly widely used as a garden plant.

This tree is one of relatively few pioneer woody species and therefore occupies open areas in or on the edges of evergreen forests and river valleys. It is fast growing with a short life-span, and may reach 15 m in height but is usually much smaller. The **bark** is greyish-brown and smooth.

Bark

At the moment two species are recognized. *V. divaricata*, which occurs in the south-eastern Cape only, is much smaller. Its leaves are glossy and dark green and the flowers are dark pink to nearly purple. On the blossom tree the **leaves** are pale greyish-green and pubescent, and *the flowers are pale purple.* (I find it virtually impossible to separate these species in the field.)

The **leaves** are imparipinnately compound with between five and 20 pairs of leaflets.

The **flowers**, which are very attractive and pea-shaped, are produced in terminal heads in spring and summer. The **pods** are velvety and are borne in pendent clusters.

Flowers

The seeds germinate very easily and, given a suitably moist habitat, seedlings grow up in their thousands and may form nearly impenetrable thickets within a few years. Under such conditions it is a very slender tree with a long, thin stem and a sparse crown. When growing alone, the crown is upright but fairly wide-spreading and dense.

52

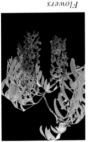

Bark

Flowers

According to information available at present, the tree wistaria extends from northern Natal to northern Transvaal and the east of Botswana, and through the eastern part of Zimbabwe to the countries north of the Zambezi. It also has a limited distribution in northern Mozambique.

Usually it is a slender, smallish tree of about 7 m, but it can grow into a much bigger tree with a spreading crown. Stems are fairly straight, often branching low. The **bark** is brownish-grey and longitudinally fissured. Although it can be found in various soil types, it seems to prefer those with a high clay content, such as those derived from dolerite. The **leaves** are imparipinnately compound with up to eight pairs of leaflets plus a terminal leaflet.

As this is one of the most attractive trees when in full flower and also one of the relatively few trees from the warmer areas of the subcontinent that can withstand low temperatures, it should be used much more freely in our gardens, with the proviso that it is not watered during the naturally dry period of the year. Although this tree is deciduous, the leaves are not dropped, or are only partially dropped, if it is watered in winter. The pendent sprays of bluish-mauve **flowers** are then partially obscured by the leaves, and the trees also seem to flower less profusely. The solution obviously lies in choosing exactly the right spot in the garden. The flowers emerge early in spring. The thin **pods** are borne in pendent clusters and bulge somewhat over each seed; they turn brown to black when ripe.

The timber resembles that of the wild olive. It is of good quality, fairly heavy and hard.

(Zimbabwe: (287) Tree wistaria)

Baphia massaiensis subsp. *obovata* (SA 223) Sand camwood

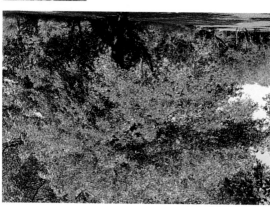

This tree is virtually an unknown entity to South Africans, as it occurs only in the northeastern corner of the Transvaal, probably only in the Kruger National Park. Further north, in eastern and northern Botswana, western Zimbabwe, the Caprivi Strip and northern Namibia, it is fairly widespread. It grows only in deep, well-drained, sandy soil. The only other member of the genus, *B. racemosa*, is endemic to the Natal coast.

The sand camwood rarely reaches 6 m in height, and is usually multi-stemmed. The  bark is smooth and dark grey to greyish-brown on young stems, becoming longitudinally ridged on old stems. It is deciduous, with a spreading crown only when growing in open woodland. The simple **leaves** are borne solitary and fairly far apart. They are obovate, up to 9 cm long, and are dull green. Two distinct swellings occur on the petiole. The most outstanding feature of this plant is its attractive, pea-like **flowers**. The petals are pure white, except for a yellow patch at *the base of the standard petal, crinkled, very delicate and sweetly scented. The sep-* *als are pinkish.* The flowers are borne in compact sprays near the ends of the twigs (summer months, but depending on rainfall, flowering can occur as late as April). The **fruit** comprises pods which are few-seeded, and are narrow at the base, broadening towards the tip and end in a sharp point. They are hard, shiny and brown when mature and split while on the trees to release the seeds.

The wood is hard, finely textured and dark brown but, probably due to a lack of pieces which are large enough, is not utilized. (Zimbabwe: (288) Jasmine pea)

Fruit

Bark

54

Psoralea pinnata (SA 225,8) Fountain bush

Bark

Flowers

The common name, fountain bush, aptly describes the habitat of this tree as it usually grows in wet situations in or near rivers, vleis and marshes. It is seldom abundant, although it has a wide distribution from the western Cape, all along the east coast as well as rather far inland, up to northern Natal and possibly southern Mozambique, the eastern Orange Free State, then north along the Drakensberg escarpment to the Soutpansberg as well as central Transvaal. From looking at its distribution map, it is justifiable to classify this species as frost-tolerant.

The fountain bush is probably deciduous. Full grown specimens can reach as much as 6 m in height, and will then have a single, bare trunk and a sparse, spreading crown. The **bark** is smooth and is pale brownish-grey in colour. The imparipinnately compound **leaves** are borne alternately. The leaflets are very fine and slender, sometimes as little as 0,1 cm wide, and are dark green and shiny. The leaves are borne close together at the ends of the slender, soft branches.

Although it is a graceful and attractive tree in its own right, the biggest asset of the fountain bush is its beautiful **flowers**. They are borne in the axils of the terminal leaves and are pea-shaped, *light blue to deep blue or even mauve, seldom whitish*. I have found some flowering plants at the Tsitsikamma in mid-August, but Palgrave (1983) states that generally those species in the western Cape flower in October/December, those in the eastern Cape from February to June, and those in Natal and the Transvaal during August/September. The **pods** are minute, measuring approximately 0,5 x 0,3 cm and are enclosed by the persistent calyx.

Mundulea sericea (SA 226) Cork bush

Bark

Flowers

The cork bush inhabits northern Natal, Swaziland, southern Mozambique, the entire Transvaal (except for the eastern Highveld), the northern and eastern areas of Botswana, northern Namibia and large areas of Zimbabwe. It occurs in a large variety of habitats, but seems to favour sandy soil and a rocky substrate.

It is usually encountered as a multi-stemmed shrub of about 2 m, but can grow into a single-stemmed tree of about 5 m high with a roundish, fairly dense crown. The **bark** is yellowish-grey, very soft and corky and forms longitudinal ridges which are easily broken off.

The imparipinnately compound **leaves** are borne alternately. The leaflets are fairly small (usually about 4 x 1.5 cm) and are *silvery very-green due to a layer of silky, silvery hairs*. The beautiful, pea-like **flowers** are usually mauve to purple, but those of some plants in the Pafuri area of the Kruger National Park are white. They are borne in dense sprays at the branch ends (October/January). The **pods** are fairly small, borne in pendent clusters and are covered with gold-en-brown hairs which later become grey.

The bark and the seeds contain a chemical substance called rotenone, which is widely used as fish poison. Nevertheless, the leaves are browsed by wild animals. This is another indigenous plant which should be planted more often in gardens, and judging by its distribution it should be fairly cold-resistant.

(Zimbabwe: 321) (Cork bush)

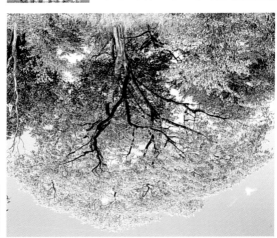

This is a beautiful, single-stemmed, deciduous tree with a wide-spreading, flattened crown. Twice a year these trees are outstandingly noticeable: when the leaves are in their full, dark-yellow autumn colour, and again in spring when the attractive orange-yellow **flowers** emerge immediately prior to, or with, the new leaves.

Wild teak is a tropical species with a preference for well-drained soil. It occurs in the northern part of Natal, Swaziland, the Transvaal Lowveld, southern and northern Mozambique, the greater part of Zimbabwe, northern Botswana and the extreme northern area of Namibia.

Bark

The **leaves** are large (up to 30 cm long), imparipinnate and pendent. The **pods** are exceptional, as they can only be recognized as such in the very early stage of development. When mature they are suborbicular, thickened and spiny in the middle, each surrounded by a thin, but fairly hard, parchment-like wing. They measure up to 10 cm in diameter, are borne in pendent clusters, and persist on the trees until the next flowering season.

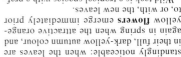
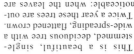

Fruit

Although the wood is fairly light, it is one of the most favoured furniture woods in the areas of Africa where it occurs. The colour of the heartwood varies from very light brown to red or even copper-brown and is very durable, easy to work with and polishes well. It is also extensively used by African artists producing sculptures of wild animals for the curio trade.

(Zimbabwe): (352) Bloodwood)

Lonchocarpus nelsii (SA 239) Kalahari apple-leaf

Flowers

Fruit

In southern Africa, this tropical species occurs only in the northern part of Namibia (including part of the Caprivi Strip), northern Botswana and predominantly the western part of Zimbabwe.

It often occurs as a multi-stemmed shrub, possibly due to veld-fires, but can be a single-stemmed tree of up to 10 m high. It grows in various habitats, but seems to thrive in dense or open woodland on Kalahari sand. Small branches often grow from the stem almost down to its base. The crown is usually dense but upright. The **bark** on young branches is smooth and yellowish-grey. It becomes dark grey and peels in thinnish flakes to expose the bark.

The **leaves** should be imparipinnately compound with one or two pairs and a terminal leaflet, but in Botswana they are usually simple. They are pale green and leathery with prominent net-veining on the underside. According to Palgrave (1983), the number of leaflets increases with higher rainfall. This is not consistent, however, and different variations may occur on the same tree. They are fairly large (up to 12 cm long), velvety when young, glabrous and glossy when old and turning bright yellow in autumn. The pea-shaped **flowers** are mauve to purple or pinkish and are borne in *large (35 cm) terminal sprays*. They appear long before the leaves (September/October). The **fruit** is a small, flat, velvety pod which is one-seeded and borne in pendent clusters.

The wood is relatively soft, light in weight and pale brown. Although not commonly used – not even as firewood – small household articles are sometimes made from it. The foliage is fair-ly heavily browsed by game.

(Zimbabwe: (359) Apple-leaf lance-pod)

Flowers

Bark

In South Africa this is a very rare tropical tree as it is restricted to the extreme north-east of the Transvaal. In Mozambique and Zimbabwe, however, it is wide-spread and covers vast areas. In Botswana it is only known to occur in the eastern and north-eastern border regions (Chobe). It also inhabits the eastern part of the Caprivi Strip.

This deciduous tree has a single tall stem and a spreading, fairly dense crown. It usually reaches a height of 10 m but may be taller. It seems to prefer well-drained soil. The **bark** on old stems is grey and peels off in small, flat, irregular blocks. The **leaves** are shed late in winter. They are large (reaching nearly 40 cm in length) and imparipinnately compound. The leaflets are big (up to 8 cm long) and dark green.

The small pea-like **flowers** are white, waxy and are borne in fairly sparse racemes at the bases of new twigs (September/December). The **fruit** comprises clusters of pendent pods which are oblong, up to 30 cm long, and flat. The central portion is thickened, prominently veined, hard and the outside is a *membranous wing*. Pods are brown at maturity and fall while still unopened. Young pods are sometimes damaged by insects; they then develop into deformed, spherical 'fruits' resembling berries. The bark exudes a blood-red cell sap when damaged, which is used as a dye. The wood is pale to dark yellow, finely textured and finishes smoothly. The leaves are browsed by game and domestic stock, although they are possibly poisonous. This tree is the host of a very rare mistletoe, *Vanwykia remota*, which, in southern Africa, has only been found in the north of the Kruger National Park.

(Zimbabwe: 360) Wing pod)

Xanthocercis zambesiaca (SA 241) Nyala tree

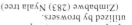

The dimensions of the only nyala tree officially measured up to now are: height 22 m; spread 29,8 m; girth 11,2 m. I measured the spread of a tree on the bank of the Limpopo River ('Tuli Block, Botswana) as 55 m. There may be even bigger ones as these are *huge, impressive trees*. It has an easterly distribution in Africa, reaching as far south as the Transvaal Lowveld. It also occurs in the north and north-western Transvaal, the east of Botswana and the southern and north-western areas of Zimbabwe as well as Mozambique.

Bark

The nyala is evergreen to semi-deciduous, usually with one short trunk. Old stems are grooved and dented: the **bark** is grey and rough but does not peel off. The branch-ends are pendent. The **leaves** are imparipinnately compound and the leaflets shiny dark green. The small, white **flowers** are borne in short, axillary or terminal sprays and are sweetly scented (November/December). Ini-

Fruit

tially the **fruit** is pod-shaped, but it soon develops into an oval or ovate drupe (berry) of up to 2,5 cm long. It becomes yellow-green to yellow-brown at maturity. The fruit pulp is floury, sticky and edible. While still on the trees, the fruit is eaten by a variety of birds as well as pri-mates. These animals also dislodge some fruit which is then eaten by nyala, bushbuck, impala and other animals. The leaves are also utilized by browsers.

(Zimbabwe (283) Nyala tree)

Erythrina lysistemon (SA 245) Common coral tree

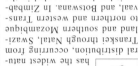

Of the six *Erythrina* species in the sub-continent, this one has the widest natural distribution, occurring from Transkei through Natal, Swaziland and southern Mozambique to northern and western Transvaal, and Botswana. In Zimbabwe it has been recorded in the south-western, central and eastern areas.

The brilliant-red **flowers** are densely packed in terminal heads which emerge long before the leaves in spring. If the trees are watered in winter, the old leaves may only partially discarded and the flowers do not show up as well as on a bare tree. They also seem to flower less profusely. The **leaves** are trifoliolate and are characterized by *small hooked thorns* on the underside of the pedicel as well as the main veins. These thorns are also scattered along the branches. The small, hard, whitish-green galls which occur in the leaves are caused by insects, and while they are not detrimental, they are unsightly. The **fruit** is a cylindrical, black pod, constricted between the seeds.

The common coral tree may be propagated by means of the seeds and cuttings. Even big branches take root easily. The tree is not very particular with regard to habitat and can therefore be grown successfully in just about all soil types and rainfall regimes.

(Zimbabwe: (366) Lucky-bean tree)

Fruit

Bark

Balanites pedicellaris (SA 252) Small green-thorn

Bark

Fruit

Although this is a widespread tropical species which extends as far south as northern Natal, it occupies only a relatively small area in southern Africa, extending from the south-eastern corner of Botswana (Tuli Block) eastwards in a narrow belt on either side of the Limpopo, into Mozambique and then southwards (also through eastern Transvaal) to north-eastern Natal.

Small green-thorn is relatively abundant in north-eastern Natal, at least in the Ndumu Game Reserve, on poorly-drained, clayey soil, where it is a component of the nearly impenetrable thornbush thickets. It is usually an untidy shrub, sometimes with long shoots, in dense bush. However, it may attain a height of 6 m in an open condition. It is then single-stemmed and upright, with a spreading, very sparse crown. Even young twigs are exceptionally hard and may vary in colour from buff-green to yellowish-brown. Old stems are deeply dented and fluted, similar to those of *B. maughamii*, and the **bark** splits into small, boat-shaped sections. It is armed with single, straight, long, sharp, hairy spines.

The **leaves** are compound with *two velvety leaflets* which are markedly brownish and pubescent when young. The yellowish-green **flowers** are inconspicuous and are borne in groups of three in the leaf-axils from spring to summer, depending on rainfall. The only time this tree is really conspicuous is when bearing **fruit**, which is sometimes borne in masses and is attractively deep orange when ripe. The fruit is about 2,5 cm in diameter, and consists of a single, hard stone covered with a thin, fleshy layer which is edible but not tasty.

(Zimbabwe: (382) Lesser torchwood)

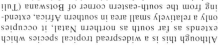

Calodendrum capense (SA 256) Cape chestnut

Flowers

Fruit

Bark

When in flower, this is one of the most beautiful trees in southern Africa. It is a forest species, occurring all along the east coast of South Africa roughly from Mossel Bay to northern Natal, and along the escarpment of Swaziland to the Soutpansberg. In Zimbabwe it inhabits mainly the eastern forests, where it also penetrates Mozambique. When travelling on the Garden Route in the south-eastern Cape during their flowering time (October/December), they are highly visible, especially when taking the old road through the Grootrivier and Bloukrans passes.

Although usually a smallish tree, it can reach 20 m in favourable conditions. It is single-stemmed, often branching fairly low down the trunk, and has a dense crown which spreads in open conditions only. The bark is grey and smooth. Although evergreen, it may lose its leaves in dry conditions. The **leaves** are simple, rather big (up to 12 x 7 cm) and occur in opposite pairs. They are aromatic and characterized by *scattered, translucent gland-dots.* The **flowers** are striking. The petals are mostly pink, sometimes nearly white, and the five sterile stamens always *pink with maroon to purple gland-dots.* They remain attractive for several weeks. The fruit is a five-lobed, woody capsule of about 4 cm in diameter, covered with small wart-like knobs.

(Zimbabwe: (391) Cape chestnut)

Vepris lanceolata (SA 261) White ironwood

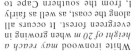

White ironwood **may reach a height of 20 m** when growing in evergreen forest. It occurs all along the coast, as well as fairly deep inland, from the southern Cape to northern Natal and the Mozambique coast as far as Beira. In the Transvaal it is confined to the mountainous regions from the Magaliesberg to the Soutpansberg. It grows mostly in dry shrub forest and in dense thickets on coastal dunes.

Under favourable conditions, the crown of this evergreen tree may be fairly wide-spreading and the trunk long and straight. The **bark** is grey to pale brown, with orange and white blotches, and smooth. The **leaves** are trifoliolate, and the leaflets are hard, dull green, markedly wavy, gland-dotted and lemon-scented.

Bark

The **flowers** are very small, yellowish and are borne in fairly large, terminal sprays (December to March, depending on rainfall). The small, roughly spherical to segmented **fruit** is borne in dense clusters and turns through purple to black at maturity (autumn to spring).

The wood is white, very hard, strong and even-grained, and has been used for numerous purposes.

Flowers

Kirkia wilmsii (SA 269) Mountain seringa

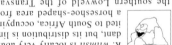

K. wilmsii is locally very abundant, but its distribution is limited to South Africa, occupying a horseshoe-shaped area from the southern Lowveld of the Transvaal northwards to the Soutpansberg and then southwards, stopping short of Pretoria and Rustenburg. It only occurs on mountains.

It is a deciduous tree which is striking and noticeable in its spring apparel of pale green as well as in the *dark reddish-brown to yellow* of autumn. It is usually either multi-stemmed or low branching. The **bark**  is dark grey and smooth with irregular sections of dead bark. The crowns of old trees have a fairly wide spread and the branches are often on the ground. Young trees are slender and upright.

Bark

The **leaves** are borne close together at the ends of the branches. They are imparipinnately compound, with a large number of leaflets, and look feathery. The margins of the leaflets are inconspicuously but coarsely serrate. The small, white **flowers** are borne in axillary panicles with long, slender stalks (September/October). The small, capsule-like **fruit** is borne in fairly large clusters. It is roughly oblong, up to 1.1 x 0.6 cm, and composed of four joined triangular parts, with the joints showing as sharp ridges. When mature, it splits along the ridges and often persists on the tree until the next flowering season.

Flowers

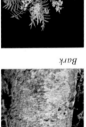

The timber is dirty white to pale-brown and fairly light and soft, yet has a fine texture. It works easily but is apparently not worth much. It has been reported that the tree is browsed by game and domestic stock.

Commiphora marlothii (SA 278) Paperbark commiphora

Bark

Fruit

The paperbark com-
miphora occurs in
central and northern
Transvaal, eastern
Botswana, various areas in Zimbabwe as well as in Mozambique,
almost always in rocky situations. It has a single, bare trunk
and a spreading, not particularly
dense, crown and is deciduous.
The stem is very characteristic. The dirty-
white to yellow **bark** peels off in *large,
thin, papery strips and flakes* – even on
young branches – and the exposed, living
bark is *bright yellow-green*. The **leaves** are
imparipinnately compound and are borne at
the tips of the branches. They may be up to
20 cm long. The leaflets are fairly large (8 x
4 cm), pale green and distinctly hairy. The
margins are largely crenate.

The small, yellow-green **flowers** are borne
in very small, dense inflorescences on long,
hairy stems (September/October, depending
on rain). The **fruit** is roughly oblong, borne
in tight, pendent clusters on long, hairy stalks. It is symmetrical
and is longitudinally subdivided by a distinct groove. The fruit
turns pale red when ripening. The seeds (stones) persist on the tree
after the flesh has fallen. The red pseudo-aril at the base covers
only a small portion of the stone, extending into four thin, smooth
'fingers'. The timber is pale brown, soft and worthless.
(Zimbabwe: (412) Paperbark commiphora)

Commiphora mossambicensis (SA 281)
Pepper-leaved commiphora

This species stops just short of the northern South African border but covers most of the Zimbabwean territory, the north and east of Botswana, the Caprivi Strip as well as the northern tip of Mozambique after which it has been named.

It may reach 10 m in height but is usually not more than about 6 m. It is single-stemmed with a sparsely branched, spreading crown. Young twigs are reddish-brown. The **bark** is brown or reddish-grey and smooth. Young sideshoots are often exceptionally thick at the base.

Bark

The **leaves** are characteristic, either trifoliolate, with two pairs and a terminal leaflet, or with one pair plus a solitary leaflet and the terminal leaflet. The leaflets are exceptionally large, ovate to almost circular (up to 8 x 8 cm) *and bright pale green*. The margins, especially in young leaves, are hair-fringed. The small, yellowish **flowers** are borne in axillary groups (October/December). The **fruit** is almost spherical, about 1 cm in diameter, and turns reddish when mature. The seed is black and the bright red pseudo-aril is very prominent, the 'fingers' nearly reaching the top of the seed. The roundish fruit sometimes takes nearly a full year to reach maturity.

Fruit

These trees are most noticeable during autumn (April/May) when the leaves turn bright yellow before being dropped. The wood is used to make small household articles, but is soft and not worth much. The leaves have a peppery smell.

(Zimbabwe: (415) Pepper-leaved commiphora)

This is a tropical species, extending southwards into Mozambique, Zimbabwe (wide-spread), north-eastern Botswana, the northern and north-eastern areas of the Transvaal as well as Swaziland and northern Natal.

It is a deciduous tree and very conspicuous when in its full autumn dress of bright yellow leaves. It can reach a height of 30 m and has a long, thick-set, straight trunk and a very wide-spreading crown. The **bark** pattern is similar to that of the marula (*Sclerocarya birrea* subsp. *caffra*). The **leaves** are paripinnately compound, up to nearly 30 cm long and are closely packed at the twig terminals. The **flowers** are inconspicuous, greenish-yellow and are borne immediately below or in the axils of the new leaves (September). In contrast to the flowers, the **fruit** is large and very conspicuous, especially when ripe (up to 22 cm long). It is pendent, cudgel-shaped, hard and woody. From an initial shiny-green, it turns purplish-brown and then dark-brown. At maturity, the pericarp splits into five valves which curve backwards, revealing the central column with the winged seeds neatly embedded in it. It resembles a peeled banana at this stage.

The wood is fairly hard, heavy and dark reddish-brown. It is held in high esteem as a furniture wood. Palgrave (1983) states that it was the royal tree of Barotseland in Zambia, and that barges were made from it for the Paramount Chiefs.

(Zimbabwe): (423) Wooden banana

Fruit

Bark

Entandrophragma caudatum (SA 293) Mountain mahogany

The Cape ash is widespread along the east coast from the Cape Peninsula to northern Natal, the Transvaal Lowveld and northern Transvaal, Mozambique and eastern Zimbabwe. An isolated population also occurs in northern Botswana. It is a tropical species which occurs as far north as Ethiopia and the Sudan.

It is a single-stemmed, evergreen tree with a dense, spreading crown and can be up to 20 m in height. The ends of its branches are pendent and are often as low as ground level. Stems are usually rather short, and the **bark** is dented, grooved, dark grey and slightly rough. The **leaves** are large (up to 30 cm long) and are borne at the ends of the twigs. They are imparipinnately compound, usually with five pairs and a terminal leaflet. The leaflets are symmetrical, between 5 and 9,5 cm long, glabrous, moderately hard, shiny and dark green.

The very small, white **flowers** are borne in long (up to 17 cm), sparse, branched racemes just below or in the axils of the older leaves (October/November). The **fruit** is roughly spherical, up to 2 cm in diameter, and attractively **bright red when ripe**. A thin, soft exocarp (outer layer) encloses a white, slightly sticky, soft fruit pulp which contains two to four seeds.

The timber is off-white to pale brown with a reddish tinge. It is finely textured, fairly light, quite soft and not durable. A variety of wild animals eat the fruit. This is one of the fastest growing indigenous trees; seeds should be planted fresh.

Flowers

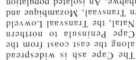

Bark

(Zimbabwe: (432) Dog plum)

Trichilia emetica (SA 301) Natal mahogany

Fruit

Flowers

This tropical species occurs from Natal to the northern areas of Zimbabwe and Mozambique. It also features in the area around Kazangula, where Namibia, Botswana and Zimbabwe meet. It grows mostly near water, or in high rainfall areas. Natal mahogany is a fairly large (up to 20 m), attractive tree with a short, often dented, trunk and a very dense and wide-spreading crown. Branch-ends are usually pendent and may reach ground level. The **bark** is grey-brown and usually smooth.

The imparipinnately compound **leaves** may

Bark

be as much as 50 cm long with the leaflets usually about half the size. They are borne quite far apart. Leaflets are elliptic to oblong-elliptic, glabrous, glossy and very dark green above and dull green with pale brown pubescence below.

The small **flowers** are tubiform, predominantly green and borne in *large, dense racemes* near the twig terminals (September/October). Some trees always bear lots of fruit while others do not. The **fruit** is pubescent, pale buff-green, roughly pear-shaped and longitudinally segmented in three capsules. It dehisces while on the tree to expose the attractive seeds, which are nearly totally enclosed by orange-coloured pseudo-arils. The small, oval portion, which is not covered, is pitch-black.

(Zimbabwe: (435) Natal mahogany)

Bark

Fruit

This tropical species extends southwards to fairly near to Pretoria near Rustenburg in the Transvaal. It is also widespread in Zimbabwe and Mozambique, parts of Botswana and the northern region of Namibia. It occurs in woodland, probably always in sandy soil, and is nowhere abundant.

Flowers

The violet tree is a single-stemmed, deciduous tree with a dense, twiggy, but poorly spreading crown. The largest specimen on which information has been published by the Dendrological Society grows in the Potgietersrus district of the Transvaal. It is 9 m high with a spread of 10,6 m. The trunk has a diameter of 0,95 m.

The **bark** is largely smooth and very pale grey and the trunk often shallowly fluted.

The beautiful **flowers** are produced in such profusion (September/November) that the trees are completely covered, given ideal climatic conditions. Typical of legume flowers in structure, they vary in colour from pink to mauve. The **fruit** is very characteristic since it consists of a roughly oval, hard, fairly small nut with a large (up to 4 cm), *membranous wing which is shaped like a hatchet*. They are purplish-green when young and light-brown when mature. They remain on the tree for about a year. The **leaves** are fairly small and are grouped together on short lateral shoots.

The fleshy roots of this plant are used, medicinally and otherwise, by African people. An oil extracted from the roots contains 99% pure methyl salicylate. It therefore has a pungent smell exactly like that of household remedies for stiff or strained muscles.

(Zimbabwe: (442) Violet tree)

This species is common in the Transvaal Lowveld and adjacent areas, but less so in the north-western Transvaal and adjoining Botswana region, and then again it is very prominent in the north of Botswana as well as the north-western and northern parts of Zimbabwe. It also features in central and north-eastern Zimbabwe and large areas of Mozambique.

It is deciduous or semi-deciduous, and reportedly can attain a height of 20 m but it is usually less than half that size, with a single, fairly short trunk and a spreading, leafy and *exceptionally dense crown*. The stem is usually low-branching, bent and grooved and fairly smooth. The **bark** is grey-brown with a yellowish tinge, with a thin outer layer of dead bark which splits into longitudinal strips.

The simple **leaves** are fairly large, usually about 7 x 5.5 cm and sometimes up to 18 x 14.5 cm, *cordate* (heart-shaped), very thin and smooth above but rough beneath. Young leaves have stellate (star-shaped) hairs on both surfaces. The margins are irregularly serrate. The inconspicuous, yellowish-green **flowers** are axillary, borne in long, sparse, spicate racemes (up to 11 cm long) at the ends of the twigs (October/November). Male and female florets occur in the same raceme, with female florets at the base. The **fruit** is roughly pear-shaped, trilocular and about 3.5 cm in length. It becomes brownish-yellow when ripe (February/March).

Flowers

Fruit

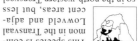

(Zimbabwe: (480) Fever-berry croton)

This tree, previously named *Ricinodendron rautanenii*, occupies a widish strip either side of the Botswana/Zimbabwe border as far as Kasane, then spreads west through northern Botswana and northern Namibia. From Kasane it extends eastwards for some distance, then stops. In South Africa it occurs only in the north-western Transvaal on the border with Botswana.

The manketti is one of the larger African trees (up to 20 m). It is deciduous and has a single, long, very thickset trunk and a wide-spreading crown. The tree in the photograph is 15 m high and the stem has a girth of 4 m. This species prefers a well-drained habitat such as Kalahari sand, and often forms pure stands. Beautiful specimens can be seen on the road between Ngoma bridge and Katima Mulilo in the Caprivi Strip.

Fruit

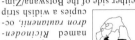

Bark

The hairy **leaves** are digitately compound with five to seven largish leaflets (up to 11 x 5 cm). The petioles are long (15 cm). The yellow **flowers** are about 10 cm in diameter, with the sexes on separate trees (October/November). The **fruit** is slightly oval, about 3 cm long and grey-green as a result of the velvety hairs covering it. The pulp, only a few millimetres thick, covers a large kernel. It is woody and difficult to crack. The seed is edible and contains yellow oil. The fruit is yellowish when ripe. (Zimbabwe: (511) Mankketi tree)

Euphorbia ingens (SA 351) Common tree euphorbia

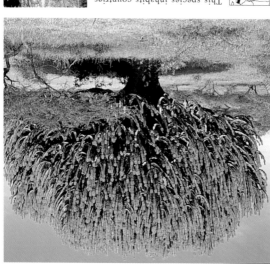

This species inhabits countries on the eastern side of Africa only. In Transkei, Swaziland, Natal, and south-eastern Transvaal, and to the north of the escarpment only. In the north of the Transvaal its range extends westwards to include eastern Botswana and to the south-west to include the Magaliesberg. In Zimbabwe it occurs in the southern, western, central and northern areas, extending into Mozambique.

The species name *ingens* means 'large, massive or enormous', which aptly describes the tree. Its stem is short and the massive crown is very dense. It can reach roughly 12 m. The **bark** is grey to dark grey, slightly rough and dented. The branches are usually four-sided with four ridges (sometimes five or six). They branch freely and the branches are conspicuously articulated and about 10 cm in diameter between opposite ridges. If the green outer layer is pierced, latex will stream from the wound. The floral buds are situated close to the spines and develop into groups of three yellow-green **flowers** on a common peduncle about 1 cm in length (May/June). Usually only one fruit develops from a group of flowers. The mature **fruit** is green, nearly spherical, up to 1,3 cm in diameter and dehisces while on the tree to release the seeds. The timber is fibrous, soft and useless.

(Zimbabwe: (527) Candelabra tree)

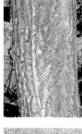

Bark

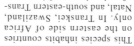

Flowers

Sclerocarya birrea subsp. *caffra* (SA 360) Marula

Flowers

Fruit

Bark

The marula's biggest asset is its **fruit** (*fairly large, yellow berries with a nauseatingly sweet smell*). The white, slimy fruit pulp is rich in vitamin C and can be eaten as is or made into a refreshing drink, a potent alcoholic beverage, an excellent conserve and the well-known pale yellow to reddish jelly. The seeds taste like walnuts and contain about 50% oil which is rich in protein. Water can be obtained from the roots. The **bark** contains a substance similar to antihistamine and has been successfully used in the treatment of blisters caused by hairy caterpillars. The timber is tough and can be used for various purposes. The leaves are utilized by a variety of game.

The marula is widespread from Natal through Swaziland into the Transvaal, Mozambique, Zimbabwe, large areas of Botswana and the northern areas of Namibia. It is a deciduous, single-stemmed tree with a wide-spreading crown.

The **leaves** are imparipinnately compound, up to 30 cm long and are crowded at the ends of the twigs. The small **flowers** are rather inconspicuous; the two sexes occur on separate trees. Flowering starts in August/September and the berries usually ripen in January/February.

(Zimbabwe): (537) Marula

75

Harpephyllum caffrum (SA 361) Wild plum

This is almost ex-
clusively a South
African tree, occur-
ring from the eastern
Cape to northern Natal, the
southernmost area of Mozambique, Swaziland and then along
the escarpment to the Soutpansberg. It is the only representative
of the genus. It is often found in
moist situations but in the eastern Cape it
also grows on rocky hillsides.

Flowers

The wild plum is evergreen, single-
stemmed and has a dense, wide-spreading
crown. The stem branches fairly low and
branch ends are sometimes pendent. It may
reach 15 m but is normally much smaller.
Its **bark** is dark brown and rough.
The oldest **leaves** are shed continuously
and *become red before dropping*. The leaves
are crowded at the twig terminals and are
imparipinnately compound with four to
eight pairs of leaflets plus a terminal leaflet.
The leaflets are slightly sickle-shaped,
asymmetric, glossy and dark green. The small, white **Flowers** are
borne in axillary sprays, the two sexes on separate trees. The oval
fruit is up to 2 cm long and is red when ripe. It is edible but sour,
and the pulp is rather thin. It can be used to produce wine and
jelly. Birds and primates relish the fruit.

Fruit

The wood is reddish, rather heavy and finishes well. It has been
used for the manufacturing of furniture and other commodities.
The wild plum has been cultivated rather extensively in both South
Africa and Zimbabwe, often as street trees, and young plants are
on sale at some nurseries.
Although not related, this tree and the Cape ash look alike. The
major difference is the symmetry/asymmetry of the leaflets.

Protorhus longifolia (SA 364) Red beech

Only its presence in Swaziland robs this species from South African endemism status. It occurs in forests and open woodland, sometimes on rocky hillsides, all along the eastern coast from the eastern Cape to northern Natal where it deviates inland and follows the escarpment to the Soutpansberg.

Red beech is an evergreen tree, with a fairly long, bare trunk and bare branches. It may reach 15 m in height. The crown is wide-spreading. The **bark** is rather smooth, grey and mottled on young branches, and cracks lengthwise with age.

The simple **leaves** are crowded near the branch tips and are borne opposite or subopposite. They are narrowly-elliptic, up to 15 cm long, hard, dark green and glossy with prominent veins, especially on the under surface. Old leaves turn yellow to bright red before being shed.

Although the yellowish-green **flowers** are very small, they are borne in fairly *large, dense, conspicuous, axillary clusters* at the twig terminals, with the two sexes on separate trees. The flowering period is from August to October. The fleshy, asymmetric **fruit** is about 1 cm in diameter, glossy and turns purple when ripening (October/December).

The bark exudes a sticky gum. The timber is of fairly good quality but is not durable.

Unripe fruit

Bark

Ripening fruit

77

Loxostylis alata (SA 365) Tarwood

Tarwood is endemic to South Africa, occurring in the eastern Cape, Natal, Ciskei and Transkei. Specimens can be seen in the Suurberg National Park as well as in the Oribi Gorge Nature Reserve. It seems as if the species prefers a fairly well-drained soil as it is often found on rocky hillsides and outcrops, but it also grows along riverbanks.

Flowers

This evergreen tree usually has a short, single stem and a wide-spreading, roundish and very dense crown. It can be up to 10 m high. The **bark** is grey to dark-grey, fairly rough and forms narrow, longitudinal ridges which sometimes crack crosswise to form small blocks.

The **leaves** are characteristic. They are imparipinnately compound with up to five pairs of leaflets, and the rachis is distinctly *winged*. Old leaves are dull green to yellowish-green, but the young leaves are often attractively tinged with red. The smallish, white, star-shaped **flowers** are borne in large, branched, terminal panicles, with the sexes on separate trees. The persistent sepals are the most conspicuous part of the flower. They are petal-like and become pink to red and eventually straw-coloured when the fruit ripens. The **fruit** is small and fleshy and is situated at the base in the middle of the star-shaped 'flower' formed by the sepals. Flowering usually starts in September/October but may be as late as the following autumn, and because of the persistent, colourful sepals the female trees look spectacular for months on end.

This species can be found in most of the bushveld areas of Natal, the Transvaal, southern Mozambique, southern Zimbabwe and large parts of Botswana as well as northern and southern Namibia. It prefers well-drained situations and is therefore found most frequently on sandy soil or hillsides. It is a deciduous tree which loses its leaves late in spring. It is normally smallish (9 m) with a short trunk and poorly spreading crown. Suckers often sprout from the base of the stem. The **bark** is pale to dark grey and breaks up into small blocks. The cell-sap is sticky and smells like resin.

Bark

The simple **leaves** are generally borne in verticils of three. They are oblong/oval, up to 17 cm long (mostly smaller), dark green above and silvery and tomentose (downy) underneath. The margins are minutely crenate (scalloped). The outstanding characteristic of the leaves is the *herringbone pattern formed by the main and secondary veins*, which is the common factor linking the *Ozoroa* species. The small, white **flowers** are borne in terminal clusters during midsummer. The striking **fruit** is kidney-shaped, initially shiny and bright green with reddish spots, becoming reddish-brown at maturity turning black and raisin-like.

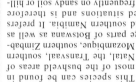

Flowers

The wood does not burn readily, probably due to the resin, and after veldfires partially-burnt logs are a common sight.
(Zimbabwe: (544) Resin tree)

Ozoroa paniculosa (SA 375) Common resin tree

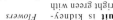

Rhus chirindensis (SA 380) Red currant rhus

This deciduous tree occurs in evergreen forests along the east coast from the south-western Cape to southern Mozambique. From northern Natal it follows the escarpment to the Soutpansberg and into Zimbabwe as far as Mutare in the east, where it also crosses into the forests on the Chimanimani mountains of Mozambique.

It is by far the largest of the wild currant trees. Large trees are common in the Tsitsikamma/Knysna forests, but outside the forests it is a smallish shrub with short, spiny-tipped lateral branchlets. Large trees have fairly long trunks and the **bark** changes from smooth and pale grey-brown to dark brown and rough with longitudinal cracks. The crown is fairly dense and wide-spreading.

The **leaves** are, like those of all other *Rhus* species, trifoliolate. The leaflets are large (up to 12 cm long) with a very long petiole, and taper into a long, prominent tip. The midrib is yellowish but is often tinged with red. The **flowers** are exceptionally small, yellowish-green and are borne in large axillary and terminal heads (October to March). The small **fruit** is round, flat and becomes red when ripe. Mature, unripe fruit has been found in August (Knysna) and nearly mature fruit in January in the National Botanical Gardens in Pretoria.

(Zimbabwe: (546) Red currant rhus)

Flowers

Fruit

Bark

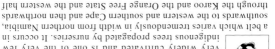

Rhus lancea (SA 386) Karree

Bark

Flowers

Fruit

As this species is one of the relatively few evergreen trees in the subcontinent which is frost-tolerant, it is very widely cultivated and is one of the very few indigenous trees propagated by nurseries. It occurs in a belt which varies tremendously in width from northern Namibia, southwards to the western and southern Cape and then northwards through the Karoo and the Orange Free State and the western half of the Transvaal into Zimbabwe and eastern Botswana. It can be found in a variety of habitats, but is usually prominent in low-lying areas near watercourses.

It generally grows to about 8 m tall, with a short stem, a dense, rather roundish crown and trailing branchlets. The **bark** is dark grey and rough. Like all *Rhus* species, it has *trifoliolate* **leaves**. They are dark green and shiny, long and the margins are entire.

The greenish, sweetly-scented **flowers** are very small and are borne in large, dense, pendent clusters at the twig-terminals, with the sexes on different trees (June/September). The small **fruit** is almost spherical and is shiny and brown when ripe.

The wood is finely textured, reddish-brown, hard, tough and durable. In areas where frost is a problem, the karree should be used to supply necessary shade.

(Zimbabwe: (550) Willow rhus)

Cassine papillosa (SA 415) Common saffron

Flowers

Fruit

The common saffron tree is mainly a South African species, following the east coast from the Cape Peninsula to Natal and then along the escarpment through Swaziland into the Transvaal as far as the Soutpansberg. It then vanishes from the scene to reappear in the 'tree-paradise' on the eastern border of Zimbabwe.

This tree frequently occurs in shrub form, but may attain a height of about 10 m in ideal conditions. It grows from sea-level to rather far inland, at forest margins as well as in wooded ravines and valleys.

It is an evergreen tree with a short trunk, branching low, and a *dense spreading, drooping crown*. The grey-brown **bark** is smooth and thin with orange underbark and small cracks.

The simple **leaves** are borne in opposite or sub-opposite pairs. They are medium-large (usually about 6 x 3 cm), leathery, thick, dark green and glossy. The margins are hardened with sharp-tipped, widely-spaced teeth. The **flowers** are inconspicuous, small, greenish-white and are borne in small, compact heads at the twig terminals (August/March). The ovoid **fruit** is white to pale lemon-yellow, up to 2,5 cm long and is often covered with wrinkles or encrustations. It reaches maturity during the following flowering season. Information on the timber is lacking.

(Zimbabwe: (576) Common saffronwood)

Bark

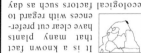

Fruit

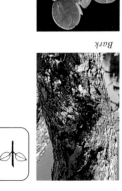

It is a known fact that many plants have clear cut preferences with regard to ecological factors such as day length, soil conditions, the availability of moisture, shelter from the wind, maximum and minimum temperatures, sun or shade, etc. Some of these plants

Leaves

are extremely sensitive to certain conditions and therefore will only grow in very small areas. For reasons unknown, *M. frangularia* is one of these as it occurs only in a restricted area in the western Cape.

The tree in the picture is one of a very small population in the West Coast National Park growing on wind-blown sand on the exposed summit of the hill called Postberg. It apparently also occurs in coastal bush and along mountain streams.

It is an evergreen tree growing to approximately 5 m in height, usually having a short, single trunk and a slightly spreading, very dense crown. The grey **bark** is somewhat rough and scaly, peeling off in flat flakes. The young **leaves** have extremely attractive colouring, being *dark red to nearly purple*. They are borne in opposite pairs and are simple, almost circular in shape, very dark green, thick, leathery and hard when old.

The **flowers** are whitish, very small and are borne near the tips of the branches. Flowering occurs from May to June. The two-seeded **fruit** occurs in small bunches, is slightly oval in shape, fairly small at 1,5 cm and has an attractive red colouring when ripe (August).

The timber is yellow, fine-grained, hard and tough and has apparently been used to make musical instruments.

Pappea capensis (SA 433) Jacket plum

This species is common in the arid southern part of Namibia and the northern Cape. In large areas of the eastern Cape it is the dominant tree and from there extends northwards into Mozambique, most of the Transvaal and Zimbabwe as well as the eastern areas of Botswana.

Fruit

Over most of its range the jacket plum is a small tree (7 m), but it can reach as much as 12 m. It is deciduous and has a short trunk and a fairly dense, spreading crown. The **leaves** are usually crowded at the ends of the branches. They are simple, variable in size, hard and rough and the margins may be entire or finely spine-toothed. The **flowers**, borne in long spikes, are male at first and later female. (P.J. Robbertse, pers. comm.) The outstanding characteristic feature of the tree is its **fruit**. It is a *softly hairy, green, round berry* up to about 1.5 cm in diameter. During some seasons the berries occur in masses. At ripening the pericarp splits in two to reveal the bright red, fleshy, shining false-aril which envelope the black seed. This fleshy part is delicious and is enjoyed by people and animals. It is also used to make jelly and to brew an alcoholic beverage.

The wood has a fine grain and is quite heavy, hard and pale brown with a reddish tinge. The quality is good enough for it to be used for various purposes but large pieces are rare. The Zimbabwean name 'Indaba tree' stems from the fact that Lobengula, a chief of the Matabele, held meetings (indabas) with his headmen in the shade of one of these trees.

(Zimbabwe: (605) Indaba tree)

Sparrmannia africana (SA 457) Cape stock-rose

As this plant is seen most frequently as a multi-stemmed shrub of 2-3 m high, its inclusion in a tree list may be criticized. However, in a suitable habitat it may reach a height of 7 m. It is endemic to the south-eastern and eastern coastal region of the Cape, as well as the Ciskei and Transkei. Palgrave (1983) refers to it as a 'rampant weed occurring on rocky hillsides and along the margins of evergreen forest'. This characteristic is well-illustrated along the Garden Route in the Tsitsikamma area, where some plants became established on the verge of a road which had been cleared during construction.

This tree should be evergreen. Its **bark** is smooth, even shiny, and greyish-brown. Its growth-habit is shrub-like, with some branches on ground level.

The simple **leaves** are large (up to 15 cm long), heart-shaped, three- to nine-veined from the base, soft and prominently hairy. They are borne alternately. The **Flowers** are most attractive and are borne in umbels in the axils of the leaves near the tips of the branches. They are very delicate and consist of four pure white petals which are folded *backwards around the pedicel,* and a central mass of stamens, some of which are sterile and golden yellow while the rest are fertile and have golden-yellow bases and maroon tips (June/November). The **fruit** is an extremely spiny, brown, dehiscent capsule.

Flowers

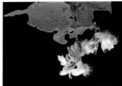

Fruit

Bark

Grewia occidentalis (SA 463) Cross-berry

Cross-berry is adapted to a divergent range of temperatures and rainfall. It occurs from the winter rainfall area of the Cape, north and north-eastwards through the evergreen forests and coastal shrub of the south-eastern Cape and the Karroid scrub of the Karoo, the scattered wooded areas in the Orange Free State, then through Natal and the Transvaal to Mozambique and the eastern part of Zimbabwe.

Flowers

It is mostly encountered as a multi-stemmed scrambler in fairly dense to dense thickets where it may reach a height of 6 m. In open woodland it is a slender tree or shrub with a fairly dense crown and trailing branchlets. Its **bark** is grey-brown and fairly smooth. The **leaves** are relatively large, rough to the touch due to coarse hairs, thin and the margins are conspicuously finely toothed. The most outstanding feature of this plant is its **flowers**. Whereas most of the Grewia species have yellow flowers and a limited number of white flowers, the minority (including this one) have *showy, pale pink to mauve, star-shaped flowers of about 3,5 cm in diameter*. Flowering extends nearly throughout the spring and summer. In accordance with the collective noun used for this group of plants (cross-berries) the **fruit** consists of four fruit lobes) joined in the centre to form a rough square or 'cross'. They are reddish-brown when ripe. The trunks are too small to be of any use, apart from firewood.
(Zimbabwe: (659) Cross-berry)

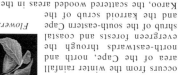

Bark

Azanza garckeana (SA 466) Snot apple

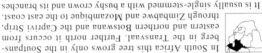

Bark	Flower	Fruit

In South Africa this tree grows only in the Soutpansberg in the Transvaal. Further north it occurs from eastern and northern Botswana and the Caprivi Strip, through Zimbabwe and Mozambique to the east coast.

It is usually single-stemmed with a bushy crown and its branches often hang to the ground. The **bark** is greyish-brown with small longitudinal ridges but fairly smooth. It seldom reaches a height of 10 m. Its **leaves** are simple, large (up to 20 x 20 cm), three- to five-lobed, with three or more major veins from the base. The upper surface is covered with coarse hairs and the underside with soft hairs. The petiole is long (up to 15 cm). The showy **flowers** are yellow, turning red when withering. Each crinkly petal has a maroon patch on the inside at the base. They are borne solitary in the axils of the leaves near the branch-ends. They flower over such a prolonged period (December/May) that the oldest fruit is nearly fully developed by the time the youngest flowers open. The **fruit** is a *characteristic, nearly round capsule* of up to 5 cm in diameter. It is longitudinally divided into five sections, covered with soft hairs and is olive-green turning to brownish-green at maturity (February to September). The tree owes its South African name to the glutinous slime produced when the fruit is chewed.

(Zimbabwe: 682) Azanza)

In height (up to 20 m) this species does not even come close to red mahogany (*Khaya nyasica*), which is said to grow to 60 m, or to the yellowwoods (*Podocarpus*) which can reach approximately 40 m, but the size of the baobab's *trunk is so great* that it qualifies for the first or second place in the world. The two largest known trees occur near Gootsapan in Botswana and Chiramba in Mozambique. The South African record baobab grows in the Letaba district: height 19 m, trunk 10,64 m in diameter.

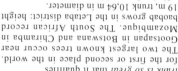

Flower

Under favourable conditions some baobabs may live in excess of 1 000 years (Guy, 1970). Initially the growth rate is extremely fast, especially between 20 and 70 years, but during the latter period of its life it slows down considerably. The trunks may even shrink during periods of severe drought. The smooth **bark**, sometimes heavily folded, is grey-brown to reddish-brown.

Bark

Because the baobab is not frost-resistant, it occurs only in the warmer, subtropical and tropical areas, including the northern area of the Transvaal as well as further north, east and in West Africa. The **fruit** was known in the herb and spice markets in Cairo as early as 2 500 BC and was known as *bu hobab*, probably derived from the Arabic words *bu hibab* meaning 'fruit with many seeds'. The large, beautiful **flowers** with their waxy, wrinkled petals are pollinated by bats. The **leaves** are palmately compound with five (seldom seven) leaflets, but are single in the case of seedlings. According to some botanists it is the only tree species representing the *Adansonia* genus in our region.

(Zimbabwe: (684) Baobab)

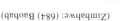

All the wild pear trees have attractive, even very showy, white to pink flowers. *D. pulchra* has the most beautiful flowers of all the species, the delicate petals being soft pink and each adorned with a red patch at the base on the inside.

Common wild pear is the most abundant and the most widespread of the species. It occurs from southern Natal northwards and covers, with few exceptions, the entire Mozambique, Zimbabwe and Transvaal area. It also features in the northern Orange Free State, eastern and northern Botswana and the north-central area of Namibia. It is often associated with rocky situations.

This deciduous tree cannot be overlooked in late winter/early spring, when it is one of the first to start flowering. It is usually covered in white, and occasionally delicate pink. **Flowers.** The **fruit** is a small, yellowish-brown, hairy nutlet situated in the centre of the base of the cup formed by the dead, light brown, persistent petals.

It is usually a single-stemmed, small (6 m), slender tree with a moderately-spreading, sparse crown. Old stems are dark grey and often bear a few lateral twigs. The bark breaks up into small, irregular blocks. The **leaves** are nearly

Bark

Flowers

circular, fairly large (up to 15 cm in diameter), *seven-veined from the base, and rough.* The margin may be entire to markedly, but irregularly, toothed.

(Zimbabwe: 687)

Dombeya rotundifolia (SA 471) Common wild pear

Sterculia africana (SA 474) African star-chestnut

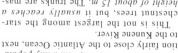

This tree occurs from the Indian Ocean all along the Zambezi to the Caprivi Strip and Chobe area of Botswana. Isolated populations also exist in eastern Botswana, south-western Zimbabwe, and the north-central area of Namibia, with a tiny population fairly close to the Atlantic Ocean, next to the Kunene River.

This is not the largest among the star-chestnut trees, but it *usually reaches a height of about 15 m*. The trunks are massive and smooth, and the colour varies from nearly white to reddish-brown. The outer **bark** flakes to reveal beautiful, pastel-coloured, marbled under-bark. The large **leaves** are three- to five-lobed, up to 15 x 13 cm, with five to seven large veins from the base, and are conspicuously hairy. The cup-shaped **flowers** are greenish-yellow with red lines, up to 2,5 cm in diameter and are borne in compact, terminal racemes. The sexes are borne apart on the same tree, and they appear in spring, before the leaves. The impressive **fruit** consists of one to five swollen carpels up to 15 cm long, each with a prominent tip. It splits on one side to release the seeds (March/April). (Zimbabwe: (689) Tick tree)

Fruit

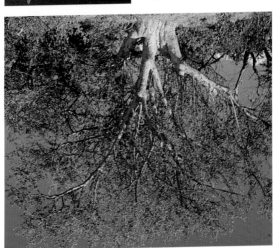

S. rogersii is endemic to southern Africa. It is fairly abundant over its entire range, which roughly encompasses the area from the east of Botswana to the Mozambique coast, and from Harare in the north through the eastern Transvaal to northern Natal. It prefers well-drained situations and is therefore found mostly on rocky outcrops.

It is a small (usually about 6 m), deciduous tree with an *exceptionally thickset, succulent-like trunk which always subdivides low down*. The crown is very sparse and spreads only moderately. It can sometimes be mistaken for a young baobab. The trunks are smooth, reddish-brown and mottled with yellow or yellowish-green patches caused by the peeling **bark**. The simple **leaves** are borne on new shoots only. They are small, usually three-lobed and cordate (heart-shaped) to broadly ovate with a distinct cordate base.

The small, cup-shaped **flowers** are borne on short lateral shoots and on the branches. They are reddish-green on the outside and yellowish-green with red, vertical lines on the inside. The **fruit** is a swollen carpel up to 8 cm in length and covered with golden hairs. Usually three to five develop to maturity. The seeds are oval, smooth and dull leaden-grey. Flowers and/or fruit can be found nearly all year round.

Leaves

Flowers and fruit

(Zimbabwe: 692) Squat sterculia

Ochna pulchra (SA 483) Peeling plane

This species occurs from the Magaliesberg range north and north-eastwards into Zimbabwe as well as into the south-east of Botswana. It is also abundant in northern Botswana and north-eastern Namibia. It is a single-stemmed, deciduous tree with a moderately dense, roundish crown. Usually between 6 and 7 m high, it mostly occurs as a shrub in dense thickets. Twigs and branches are extremely brittle and nearly white. Old stems are smooth, cream-coloured and usually almost completely covered by thin, hard, curled flakes of peeling bark. Very old stems are dark grey and the **bark** peels in small, flat sections.

The simple **leaves** are borne at the twig terminals. New leaves are very glossy and can be pale green, brownish-green or even red-brown. Mature leaves may be up to 10 cm in length and are glossy, hard and brittle. Leaf margins are usually entire but sometimes finely serrated in the upper third. The **flowers** are up to 2 cm in diameter, pale yellow and borne in masses (August/October). The **fruit** is borne in pendent clusters and changes colour from pale or olive-green to pitch-black. (Zimbabwe: (706) Peeling-bark ochna)

Flowers

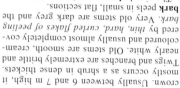

Fruit

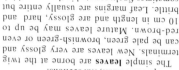

Garcinia livingstonei (SA 486) Lowveld mangosteen

G. livingstonei occurs from
northern Natal northwards
through Mozambique, Swaziland and the Transvaal Lowveld
to south-eastern Zimbabwe. It reappears
along the Zambezi River from Mozambique
to the Caprivi Strip, north-western Zimbabwe and northern Botswana.

It is a handsome, distinctive, evergreen
tree (up to 12 m). Although single-stemmed,
it often splits into two or more secondary
stems low down. Old **bark** is grey to black
and subdivided into small, regular sections.
The branches are rigid and in full-grown
trees the crown is invariably topped by some *long shoots sticking
out above the canopy*. The species contains yellow latex which is
also exuded by the veins when leaves are cracked across.

Leaves occur in verticils of three and persist even on fairly thick
branches. They are simple, red and soft when young but hard, dark
green, thick and brittle when mature. The veins are clearly visible
since they are yellow-green. The small **flowers** are pale green to
yellowy-green and are borne in small groups in the leaf axils on
older branchlets (September/November). Male and bisexual flowers are borne on separate trees. As can be seen in the illustration,
the **fruit** is very striking, up to about 3,5 cm in diameter, and is
produced in profusion. It is edible and is enjoyed by people and
animals alike. An alcoholic beverage is also prepared from it.
(Zimbabwe: (716) African mangosteen)

Fruit

Monotes glaber (Zim. 718) Pale-fruited monotes

The *Monotes* genus is the only one in the Dipterocarpaceae (meaning 'winged fruit') family represented in the tree flora of the subcontinent of southern Africa. Apart from Zimbabwe, *M. glaber* can only be seen in the east and north-east of Botswana. I found it a few kilometres south of Kazangula in Botswana, which is far north of its previously known distribution range in that country, and more fieldwork may reveal its existence in the Caprivi Strip as well. It grows in fairly dense to open woodland on sandy soil.

This is a fair-sized tree. The one in the picture is 13 m high with a long, bare trunk and a dense, widespreading crown. It is probably deciduous. The **bark** is greybrown and fairly smooth.

Bark

The simple **leaves** are oblong to elliptic, up to 10 cm long (usually 7 cm), glossy and light green above and the yellowish veins are conspicuous. A characteristic and *distinctive spot (nectary) occurs at the base of the midrib*. The pale greenish-yellow **flowers** are star-like and are borne in lax, relatively few-flowered axillary heads (2-4 cm long), which appear between November and March. The **fruit** is slightly oval, ridged, about 1 cm in diameter and is situated in the centre of the five persistent calyx lobes which *are wing-like and straw-coloured when ripe*, giving the impression of a wooden star-like flower. They are up to 3 cm long.

Fruit

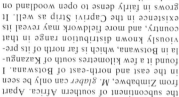

The timber is light brown with darker striations and is suitable for the manufacture of furniture.

Oncoba spinosa (SA 492) Snuff-box tree

This is a tropical species occurring as far south as Natal, but nowhere in South Africa is it really abundant. It usually only attains shrub size and is therefore an unknown entity among local dendrologists. In Zimbabwe it tends to form thickets. Available information points to its presence in Swaziland, Mozambique, the Transvaal Lowveld and northern Transvaal as well as all along the Zambezi into a limited area of north-eastern Botswana and the Caprivi district of Namibia. Judging from its habitat preference in the Transvaal Lowveld, it prefers moist conditions such as river banks.

The snuff-box tree is semi-deciduous to deciduous, usually multi-stemmed with a dense crown and up to 6 m in height. The **bark** is greyish-brown and rather smooth. It is armed with fairly long, thin, solitary spines. The **leaves** are big (up to 12 x 6 cm – usually somewhat smaller), simple, thin leathery and ovate.

The most outstanding asset of this tree is its beautiful, sweetly-scented **flowers**. They may be up to 10 cm in diameter with a large number of pure white, crinkled petals and a central mass of yellow stamens – *resembling a fried egg*. They are borne solitary and axillary or terminally (September/January). The **fruit** is spherical, hard-shelled, up to 6 cm in diameter, indehiscent and brown when mature. The brown seeds are embedded in a dry, yellow, edible pulp.

The wood is light brown and hard but is seldom used as large pieces are not available. When the fruit is dry, the seeds rattle inside it and this is used to amuse small children as well as for attaching to anklets and bracelets for dancers. They are also used as snuff-boxes.

(Zimbabwe: (726) Fried-egg flower)

Fruit

Flower

Kiggelaria africana (SA 494) Wild peach

This is one of the larger trees, attaining a height of up to 27 m when it grows in the evergreen forests from the southern Cape to the Soutpansberg, as well as the Mutare region of Zimbabwe and adjoining Mozambique territory. However, in the more arid regions of Namaqualand, the western Cape and the Karoo as well as the Orange Free State, Lesotho, the interior of Natal and the Transvaal and south-eastern Botswana it is much smaller.

Fruit

It is a single-stemmed tree with a wide-spreading crown in suitable habitats. The trunk is straight and sometimes very long; the **bark** is smooth and pale brown when young, becoming dark brown and flaky when old. The simple **leaves** are oblong to elliptic, up to 9 x 5 cm but more often smaller, and softly hairy when young but later glabrous. The margins may be entire or markedly serrate, sometimes in the upper half only. Small warts occur in the axils of the secondary veins.

Bark

Male flowers, borne in sparse heads, and female flowers, borne solitary, occur on separate trees. They are 1 cm in diameter and yellowish-green to greenish-white (August/January). The **fruit** is a *characteristic spherical capsule* up to 2 cm in diameter. It is rough and greyish-green, splitting into four valves at maturity. The black seeds are covered by an orange-red, sticky coating. The fruit is eaten by birds.

(Zimbabwe): (728) Pink-wood)

Olinia emarginata (SA 514) Mountain hard pear

In southern Africa this genus is represented by five tree species. They share one rather extraordinary characteristic: all have an extremely limited distribution. Four are South African (although *O. rochetiana* is a tropical species) and occur largely on the eastern side of the country. The fifth is limited to eastern Zimbabwe, and possibly Mozambique.

O. emarginata has the widest distribution. It extends from the eastern Cape through Ciskei, Transkei, Lesotho, Natal, the eastern Orange Free State and Swaziland to the bushveld areas of the Transvaal north of the Magaliesberg, but not reaching the Soutpansberg or the Lowveld. It is therefore very well adapted to a wide range of varying ecological conditions, from coastal to montane.

In a suitable habitat like the evergreen forests, it is a large tree (up to 20 m) with a single stem and a dense, wide-spreading crown. The **bark** is light grey on young branches, becoming dark grey and finely, longitudinally ridged when older. The simple **leaves** are borne in opposite pairs. They are smallish (up to 5 cm long), elliptic, glossy and exceptionally dark green on the upper surface but are pale green below. The small, pink **flowers** are borne in axillary heads during the summer. The most conspicuous part of this tree is its **fruit**, which occurs in small but dense clusters and is *nearly spherical, glossy and attractively dark red when ripe* (March to June).

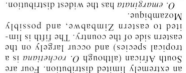

Bark

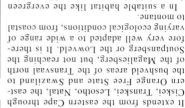

Fruit

This attractive, deciduous tree is largely confined to South Africa, in an area from the Ciskei, north along the coast as well as far inland, through Transkei, Lesotho, Natal and Swaziland into the Transvaal where it mainly occurs along the Drakensberg range up to a point near Louis Trichardt. In Zimbabwe it is restricted to a small area in the south as well as the Chimanimani area in the east, where it also crosses over into Mozambique. It mainly grows at the margins of forests or in bush on rocky hillsides.

Flowers

It is usually a very slender tree of about 5 m in height, but in favourable open conditions the crown may spread rather wide and a height of 10 to 12 m is possible. Its **bark** is smooth and brownish grey.

The **leaves** are simple, of medium size (up to 10 x 6 cm) and are arranged in precise opposite pairs. New leaves are light green and older ones bluish-green, turning yellow to brown before being shed. The greatest asset of the tree is its beautiful **flowers**. They are tubular, up to 3 cm long, *pink to pinkish-mauve and are borne in dense, nearly spherical terminal heads* up to 4 cm in diameter. The flowering period may be any time between early summer and autumn depending on the climate of the area. Flowers remain showy for about three weeks. They then die, turn grey and later black and remain on the trees for a long time. The **fruit** is a minute nutlet. The shield-like bracts at the base of the inflorescences become hard on dying and persist on the tree for over a year. High quality rope is produced from the bark.

(Zimbabwe: (760) Pompon tree)

Bark

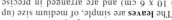

Dais cotinifolia (SA 521) Pompon tree

Combretum apiculatum subsp. *apiculatum*
(SA 532) Red bushwillow

This is a particularly common tree in the northern part of the subcontinent, occurring in all the countries involved. It occurs as far south as northern Natal in the east and central Namibia in the west. It avoids the very arid areas as well as those where winter temperatures may be low. It grows in fairly well-drained, sandy, gravelly or rocky situations only. In a favourable habitat it is often the dominant or co-dominant woody species. Red bushwillow is a deciduous, small tree with a short stem which usually subdivides low down, or it may have more than one stem from ground level. The crown has a fairly wide spread, but is sparse. It seldom surpasses 8 or 9 m in height, but I measured one of 14 m at Kubu Lodge in Botswana. Old stems are grey to greyish-black and the **bark** cracks shallowly into small, irregular sections which may peel off in thin flakes.

The simple **leaves** are usually borne in decussate pairs. They are elliptic, often obovate, and are characterized by *curved or twisted, tapering tips*. Young leaves are glossy. The small, pale yellowy-green, fragrant **florets** are borne in axillary spikes (August/November). Flower buds are often reddish-purple. The **fruit** is typically four-winged, roughly oval and on average 2,5 x 2 cm. The colour of the fruit changes from green, through red-brown to dark brown. The seed contains a chemical substance which causes severe hiccuping when eaten.

The timber is well known and claimed by many to be the best for barbecue purposes. It is extremely heavy, hard and strong. The leaves of this tree are browsed by game.

(Zimbabwe: (766) Red bushwillow)

Fruit

Combretum imberbe (SA 539) Leadwood

Flowers

Fruit

This is the largest of some 40 tree species or sub-species in the *Combretum* genus found in southern Africa. Leadwood has, with minor exceptions, the same distribution as the red bushwillow, but is nowhere as abundant and prefers soils with a high clay content.

It is a deciduous tree, most often having a single trunk and a widespreading, sparse crown. Old **bark** is pale grey to greyish-black, and cracks into small, flat, irregular blocks. The **leaves** are simple, mostly smallish (3,5 x 1,7 cm), elliptic to narrowly obo-vate, glabrous, conspicuously undulating and *grey-green due to a dense cover of silvery, microscopic scales*. They are usually arranged in decussate pairs. The small, greenish-white to yellow-ish-green **flowers** are borne in sparse, axillary panicles (Novem-ber/December). The four-winged **fruit** occurs singly or in small groups. It is nearly spherical, small (to 1,5 cm in diameter) initial-ly pale green, and buff-coloured when mature.

The timber is exceptionally heavy (approximately 1 200 kg per cubic metre) and very hard – hence the common name. It is fine-grained and dark brown to black in colour, and is virtually impos-sible to work with. Some years back, the trunk of a dead tree in Swaziland was carbon dated by the Council for Scientific and Industrial Research (CSIR) in Pretoria. It was far less than a metre in diameter at ground level but its age was determined at 1 000 years, plus or minus 50 years. It was estimated that it had been dead for about 150 years, yet it was still standing upright. (Zimbabwe: (773) Leadwood)

Combretum microphyllum (SA 545) Flame creeper

Flowers

Fruit

This is not a tree in the true sense of the word: it is a vigorous and robust climber which may reach, with the assistance of its neighbours, a height of 15 m, possibly more. It is very conspicuous and spectacular when in flower. It occurs at the margins of evergreen forests, in high rainfall areas and at low altitudes, such as the Lowveld areas of Zimbabwe, Transvaal and Swaziland, always in riverbeds where it scrambles over neighbouring trees and shrubs to form thickets. Apart from the areas mentioned, the flame creeper mainly inhabits northern Natal, large parts of Mozambique, northern and north-western Transvaal and most of Zimbabwe.

Old **bark** is pale greyish-brown and flakes in old specimens only. The simple **leaves** are borne in opposite pairs and are fairly large, glossy and dark green. They are dropped in winter. The *basal section of the petiole remains intact and develops into a blunt spine*. Emergence of the **flowers** long before the leaves is a sure sign of spring approaching. According to the local African people, it heralds the new season for tigerfish-anglers. The individual, crimson-red florets are small, but they are produced in such quantities that the trailing branches look like large sprays (August/September). The **stamens** are the most conspicuous components of the florets, the petals being very small. The typically *Combretum* **fruit** is four-winged (seldom five-), smallish (2 cm diameter), and changes colour from very light green through pink and dark red-brown to straw-coloured.

(Zimbabwe): 779) Burning-bush combretum

Pteleopsis myrtifolia (SA 547) Stink bushwillow

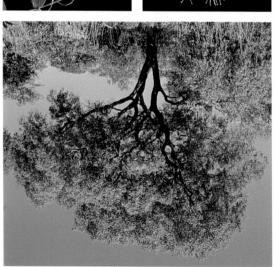

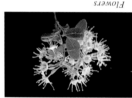

Flowers

Fruit

In South Africa this tree occurs only in Maputaland in northern Natal and in the extreme north-eastern corner of the Transvaal. In Zimbabwe it is restricted to a fairly narrow region along the border, except in the south-west. It features only in the far north-east of Botswana and the eastern extremity of the Caprivi Strip, but is widespread in Mozambique. It is found mainly in dense woodland on deep sand. It is most often encountered as a multi-stemmed shrub, but may grow into a fairly large tree (15 m) with a bare trunk and a wide-spreading crown, characterized by slender, pendent branchlets. It is deciduous. The colour of the **bark** on branches and stems varies from almost white (sunny side) to dark grey, or even black. Old bark forms longitudinal ridges.

The simple **leaves** are borne alternately, sub-opposite or oppo-site. They are rather small (usually up to 3 cm long), dark green and glabrous above and pale green and pubescent below. The small, white **flowers** are strongly and unpleasantly scented, and are borne in short, axillary heads (October to February, depending on the rainfall). The **fruit** is borne in pendent clusters. It is small (up to 2 cm long), oval to orbicular and mostly two-winged but it may have three (sometimes four) *thin, pergamenaceous wings.* (Zimbabwe: 784) Two-winged pteleopsis

Fruit

Bark

Terminalia randii (SA 550,1) Thorny cluster-leaf

In southern Africa, this is *the only Terminalia species with thorns*. It occurs in the eastern and northern areas of Botswana, the eastern tip of the Caprivi Strip and the west of Zimbabwe. It does not occur in South Africa. In the Kasane area of Botswana it grows on the dry, rocky hillsides; elsewhere, apparently, it also grows on barren, stony ground and black soils.

Flowers

The tree is medium-sized, barely reaching 10 m in height. It is single-stemmed, often branching fairly low; the crown has a poor spread and even small twigs are very rigid. It branches freely and each branch, with its twigs, tends to form layers. The **bark** is grey and longitudinally fissured. The small **leaves** are narrowly obovate and occur in clusters in the axils of the short, sharp thorns.

The small, white, sweetly-scented **Flowers** are borne in short (3 cm), sparse spikes together with the leaves (November/April). The **fruit** must be the smallest of all the *Terminalias*, being up to 2,5 x 1,2 cm, but usually smaller. It consists of the typical hard, woody fruit with a thin, pergamentaceous wing around it. When young it is purplish, but turns pale brown when dry. The fruit is borne on slender stalks which break easily, but some may remain on the tree until the following flowering season.

(Zimbabwe: (789) Small-leaved terminalia)

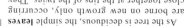

Silver terminalia is one of the most abundant trees in southern Africa: not only does it occur over very large tracts of land in all four northern countries of the subcontinent as well as in the Transvaal, northern Cape, Swaziland and northern Natal, but wherever the habitat is suitable it thrives and it is invariably the dominant tree, forming dense, often homogeneous, stands. Its preference for a well-drained, sandy soil is its outstanding characteristic.

Fruit

T. sericea is usually not more than about 8 m high. Young trees are slender. With age, the crown spreads and often tends to be flat. Stems are comparatively short, pale to dark grey with the **bark** splitting lengthwise to form ridges.

As the tree is deciduous, the simple **leaves** are borne on new growth only, occurring close together at the tips of the twigs. They are narrowly obovate, up to 10 x 2,5 cm and are, particularly when young, *covered by silky hairs which give them a silvery sheen*. The **flowers** are small (4 mm), off-white to yellowish, and are borne in fairly long spikes among the leaves. They have a nauseating smell. The **fruit** is borne in clusters. It is flat, roughly oval in shape and measures up to 3,5 cm in length. It consists of a hard, thickened inner portion with a stiff, thin, undulating wing around it. Young fruit is pubescent, but the hairs are gradually lost. Its colour changes from buff-green, through a striking pale red (sunny side) to pale brown at maturity.

Bark

(Zimbabwe: (791) Silver terminalia)

Syzygium guineense (SA 557) Water pear

This species varies greatly in different areas. It occurs on the eastern side of South Africa from Transkei northwards, including Swaziland and the western side of Mozambique, into Zimbabwe and then westwards to Botswana, the Caprivi Strip and a small population in northern Namibia.

The water pear is an evergreen tree and nearly always grows in moist conditions, sometimes even in water. It is single-stemmed and usually no more than 9 or 10 m high, but it can grow to double that size in forests. The simple **leaves** are fairly big (up to 14 cm long) and are borne in decussate pairs. The upper surface is glossy, green and pale green to darkish-green but never as dark green as those of *S. cordatum* with which it is often associated. The veins are mostly yellowish-green, but sometimes red. *The prominent yellowish petioles may be up to 3 cm long.*

The white, sweetly-scented **flowers** resemble those of the guava and some Australian blue gums, and are borne in terminal clusters. The white stamens are the most important and visible component of the flowers.

Flowers

The **fruit** is an oval berry, usually up to 2 cm long – those in the picture are 4,5 cm – and dark purple or maroon when ripe, often bicoloured on trees at Chobe. The berries are fleshy and edible and wild animals relish them. (Zimbabwe: (801) Woodland waterberry)

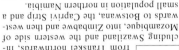

Fruit

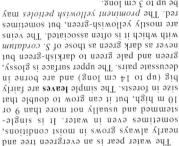

Cussonia natalensis (SA 562) Rock cabbage tree

The rock cabbage tree has a fairly wide distribution, its range encompassing Natal, Swaziland, eastern and northern Transvaal, the eastern corner of Botswana and a relatively wide strip across Zimbabwe to the north-eastern border with Mozambique. It prefers a rocky habitat.

It is a small (7 m), evergreen tree with a short, thickest trunk and a wide-spreading, rounded crown. Up to now the largest specimen measured in South Africa is 10,5 m high (Potgietersrus). The ends of the branches are similar to those of the other *Cussonia* species, being exceptionally thick, soft and succulent-like. Old trunks are grey and the corky **bark** is deeply grooved longitudinally. The ridges are broken up into rectangular sections by crosswise cracks.

Bark

The simple **leaves** are borne spirally, usually close together at the tips of the branches, and are deeply five-lobed. Occasionally some lobes do not develop fully. The whole leaf is about 15 cm in diameter. The margins of the lobes are conspicuously serrate. **Inflorescences** are borne terminally on leafless lateral shoots. When young they

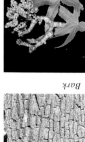

Inflorescence

resemble medieval maces. They consist of a large number of short spikes on which the sessile florets are closely packed. They have an unpleasant smell. The **fruit** is small, round, fleshy and purplish-red when ripe. The timber is soft and pale brown and is not used.

(Zimbabwe: (814) Simple-leaved cabbage tree)

Steganotaenia araliacea (SA 569) Carrot tree

Bark

Leaves

When seeing the inflorescence of this tree, anybody who is familiar with that of the carrot will undoubtedly know why it has been given this common name. Palgrave (1983) states that it is 'more characteristic of low altitude woodland, also

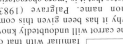

Flowers

on rocky outcrops', but in South Africa it grows only in rocky situations – outcrops as well as hillsides. It is fairly rare and is never abundant, but occurs over a wide range of territory stretching from Swaziland through the eastern and central parts of the Transvaal into eastern Botswana as well as western and southern Zimbabwe. It also inhabits an area in the vicinity of Harare and northern Botswana, and roughly the northern half of Namibia.

It is a deciduous tree with a single, short, bare trunk and an extremely sparse crown, and may reach 8 m in height but is usually smaller. Young twigs are thick, very brittle and succulent-like. The branches and stem are smooth, bright yellow-green or yellowy-grey and the **bark** peels off in thin, chartaceous flakes. On very old stems the bark is corky and horizontally cracked. The imparipinnately compound **leaves** are borne close together at the branch ends. The leaflets are a bright pale green, very thin and soft. They are characterized by dentate margins, with the 'teeth' ending in thin, soft mucros. The **flowers** are tiny, *yellowish and are borne in compound umbels* (July/November). The **fruit** is a small, flat, two-winged capsule.

The wood is not used. All parts of the tree have the familiar carrot smell.

(Zimbabwe: (817) Carrot tree)

Rapanea melanophloeos (SA 578) Cape beech

J. ROEDOLF

The Cape beech is a constituent of the evergreen forests on the eastern side of the subcontinent in the area between the coast and the escarpment, extending from Cape Town to the north-eastern Transvaal (Drakensberg) as well as the east and north of Zimbabwe, including a small portion of Mozambique. Since it is a tropical species it also features in countries north of the Zambezi. This tree may reach a height of 25 m but is normally much smaller. It is mostly single-stemmed and occurs in abundance in evergreen forests as well as the drier coastal forests.

Flowers

Young stems are whitish-pink with small, corky knobs. With age the **bark** becomes corky, longitudinally fissured and greyish-brown. Older trunks may have large, roundish knobs. In forests the crowns of young trees are long and slender, only spreading when reaching the forest canopy.

The simple **leaves** are clustered at the twig terminals and are fairly large (up to 14 x 4 cm), dark green above, paler below and the margins are rolled under. *The petioles are distinctive since they are grooved above and always coloured red to purple.* Male and female **flowers** are borne on different trees (winter to middle summer). They are small, greenish-cream and occur in masses in the leaf-axils but mostly on the older wood below the leaves. The small, spherical **fruits** often occur in such large quantities that the twigs are completely covered. They are crowded close to the branchlets and change colour from green to white and purple when ripe. Flowers and fruits are often found at the same time of year. The wood has a reticulated grain, very similar to that of members of the Proteaceae family.

(Zimbabwe: (830) Rapanea)

Fruit

108

Sideroxylon inerme (SA 579) White milkwood

White milkwood is primarily a coastal species throughout the greatest part of its range, which extends from the Cape Peninsula to the north of Natal, but from there it spreads out to include the interior of southern Mozambique, the Lowveld of the Transvaal and southern Zimbabwe.

It is a smallish evergreen tree or shrub; as a tree this species may grow to as much as 10 m in height. It is a very dense, dark green tree with a single stem, branching low, and has a *very wide-spreading crown*. Branches usually rest on the ground. The tree contains a milky latex.

Bark

The simple **leaves** are roughly elliptic, on average 8 x 3 cm. They are thick, hard, glossy and dark green. The inconspicuous greenish-white **flowers** are small and are borne either solitary or in clusters in the leaf axils or on the leafless older branches from September to April. The **fruit** is a spherical, smooth, glossy berry of roughly 1 cm in diameter. The layer of fruit pulp is relatively thick and contains latex. It turns black on ripening.

Fruit

The timber is heavy, very hard, pale brown, finely textured and durable, and has already been put to many uses. This is a protected species in South Africa and three specimens have been declared national monuments. The best known of these is the Post Office Tree in Mossel Bay, which could be more than 600 years old. The other is in Woodstock in the Cape and is called the Treaty Tree. The third is near Peddie in the eastern Cape.

(Zimbabwe: 832) (White milkwood)

This species, previously named *Bequaertiodendron magalis-montanum* occurs in all the mountainous regions of north-ern Natal and Swaziland, throughout the Transvaal as well as the southeast of Botswana. According to Palgrave (1983), they are locally common in evergreen forests, woods or ravines and along river-banks, especially among rocks in sandy soil in eastern Zimbabwe and on the Mozam-bique side of the border.

On mountains it is a smallish tree (5 m) but it can grow to at least 10 m in forests. The trunk is short and frequently crooked, and the very dark green crown is spreading and generally dense. Young twigs are con-spicuously rusty-brown due to a dense layer of hairs. The **bark** on the trunks and branch-es is fairly smooth, but is covered with small, brown to blackish protuberances, on which the flowers and fruit are borne. The simple **leaves** are borne relatively far apart and, especially in young plants, persist even on thick branchlets. They are fairly big (up to 15 x 5,6 cm), very dark green and glossy above and buff-brown and tomentose under-neath. The fragrant **Flowers** are small and *cream-coloured and appear in small clusters on the trunk and branches as well as in the leaf-axils.* The elliptic **fruit** may be up to 2,5 cm in length and is orange to deep maroon when ripe (December).

Wild animals, especially primates, relish the fruit. It is rich in vitamin C and is used to produce vinegar, wine and syrup.

(Zimbabwe: 838) Stem-fruit)

Bark

Flowers

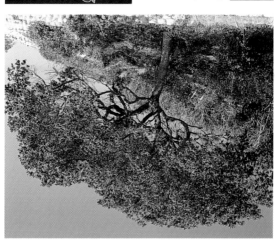

Minusops caffra (SA 583) Coastal red milkwood

As the common name implies, this tree occurs only along the coast. It is evergreen and is one of the important, sometimes dominant, constituents of the coastal dune vegetation roughly from East London to as far north as Beira in Mozambique. This is a very hardy plant which can tolerate strong winds and salt spray from the sea so well that it even occurs in thickets down to the high tide mark, as is markedly illustrated, for instance, at Cintza mouth near East London.

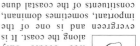

Flowers

Although it often occurs in large shrub form, it can be a large tree (15 m) with a single trunk and a very dense and wide-spreading crown. The **bark**, even on old stems, is fairly smooth and only longitudinally cracked. It is grey to dark grey with white patches.

Young branchlets and leaves are covered with *dark, rusty-brown hairs*. The **leaves** are simple, obovate-oblong, up to 7 cm long, with a prominent petiole; they are hard, glossy and dark green above, pale green with conspicuous hairs below.

Fruit

The white, star-shaped **flowers** are borne in small groups in the axils of the terminal leaves. The sepals are covered by rusty-brown hairs. The tasty, edible **fruit** is an oval berry of up to 2 cm in length, becoming attractively orange-red when ripe.

The reddish timber is finely grained, heavy, hard and durable. It is used for boat building. As with the other *Minusops* species, it contains latex.

Euclea pseudebenus (SA 598) Ebony tree

A variety of common names are used to describe this tree, but all of them contain the common denominator 'ebony', which refers to the black heartwood. This rather unusual, graceful tree is drought-resistant and frost-tolerant, and it occurs in the very arid western areas of the subcontinent, from the north-western Cape (mainly along the Orange River, west of Upington) up to the northern border of Namibia.

Old stems are dark grey and the **bark** breaks up into fairly large, thin flakes which sometimes peel off. This tree is mostly multi-stemmed from ground level. The crown is very dense and is characterized by *long, slender, drooping branches.* It is evergreen. The **leaves** are narrow, slender (up to 5 x 0,5 cm) and slightly curved and arranged spirally. They are leathery, yellowish-green and may be softly hairy when young.

The very small, greenish **flowers** are borne in small, axillary clusters during August to September, although sometimes much later. The small (0,5 cm) spherical **fruit** is produced in masses, and remains on the tree for an extended period. It changes colour from pale green, through red to nearly black. It is edible and is eaten by wild animals, but it is not really tasty. The wood is beautiful, fine-grained and black. It is suitable for various commodities, but the trunks are often too small to be used.

Bark

Fruit

Diospyros mespiliformis (SA 606) Jackal-berry

The jackal-berry is a tropical species occurring in all the countries of southern Africa except Lesotho. In South Africa it is limited to northern Natal and the eastern side of the Transvaal and in Namibia to the north-east, which is the area with the highest rainfall. It prefers moist conditions and therefore grows most often along rivers and streams, and is often associated with termite mounds.

It is a fairly large, deciduous tree (up to 20 m) with a single, long, fluted trunk and a very dense, wide-spreading crown. Old stems are grey to black and the **bark** breaks up and peels off in small, flat sections. The simple **leaves** are initially pale green or pale brown, later dark green. They are usually oblong and up to 8 x 3 cm. The **flowers** are cream-coloured to white, tubiform, *bell-shaped* and are borne in the leaf-axis; the flowers are unisexual, borne on separate trees, and female flowers are borne singly (December). The **fruit** is a spherical berry, about 2 cm in diameter and yellow when ripe. A persistent calyx lobe partially covers it at the base. The fruit pulp is jelly-like, edible and tasty. Berries remain on the tree for a long time and are often seen on leafless trees during spring.

Dry timber is pale red with a brownish tinge. It is fairly hard and durable. It was extensively used in the wagon-building trade. (Zimbabwe: (857) Ebony diospyros)

Flowers

Fruit

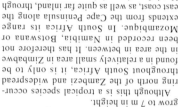

Under favourable conditions the blad-der-nut may be attractive enough to be a garden subject – particularly as it is evergreen. Although sometimes single-stemmed, it is usually multi-stemmed with a spreading, dense crown. It may grow to 7 m in height.

Fruit

Although this is a tropical species occurring north of the Zambezi and widespread throughout South Africa, it is only to be found in a relatively small area in Zimbabwe in the area in between. It has therefore not been recorded in Namibia, Botswana or Mozambique. In South Africa its range extends from the Cape Peninsula along the east coast, as well as quite far inland, through Natal, Lesotho and the eastern Orange Free State to the Transvaal where it is widespread.

The bladder-nut's **leaves** are quite distinctive: they are simple, usually ovate to oblong, up to 4,5 x 2,5 cm, very dark green,

Bark

and strikingly shiny above while pale green and hairy below. The margins are fringed with hairs. Leaves are borne alternately. The **flowers** are white to cream, fragrant and are borne in short axillary sprays (August/November), with the sexes separate. The unique **fruit** is borne singly and is roughly round, up to 2 cm in diameter, turning red when mature. It is completely enclosed by the sepals *which enlarge tremendously and become fused to form a longitudinally segmented bladder*. The sepals remain on the tree long after the fruit has fallen.

(Zimbabwe: 863) Bladder-nut)

Bark

Flowers

This is a tropical species. It occurs as far south as Natal, including Swaziland, Mozambique (south and north), the northern and eastern regions of the Transvaal as well as the eastern part of Zimbabwe.

Usually wild jasmine reaches a height of about 8 m, but in evergreen forest it may be twice as high. The only officially measured specimen in South Africa is 20 m high, but its spread is only 6,9 m. It is a deciduous tree with a rather sparse, upright but rounded crown and a single, short, grooved stem. The **bark** is grey-brown and soft, but rough. It breaks up into small, irregular sections.

The **leaves** are imparipinnately compound with two (seldom one) pairs of leaflets, and are borne in decussate pairs. The petiole and rachis are distinctive as they usually have conspicuous wings (fillodes) laterally. The beautiful, sweetly-scented **flowers** are borne in axillary and terminal cymes (September to February). They are trumpet-shaped, *about 1,5 cm long and white, pale to intensely pink and dark red at the base of the petals. The **fruit** is a hard, woody, glabrous, wedge-shaped capsule which splits lengthwise while on the tree and persists for quite a while. The flat seeds are winged.

The **wood** is pale to reddish brown, hard and durable. When the living wood is exposed through debarking, it turns bright purple and later dark brown. The timber is often damaged by woodborers and even thin twigs are often hollow.

(Zimbabwe: (864) Wing-leaved wooden-pear)

Olea europaea subsp. *africana* (SA 617) Wild olive

The wild olive is drought-resistant and frost-tolerant, and is one of only a few tree species to be found almost throughout South Africa. It also occurs in southern Mozambique, the western part of Namibia and fairly large areas of Zimbabwe. It is rare in Botswana and apparently limited to the south-eastern sections.

It is a single-stemmed, evergreen tree with a moderately spreading crown. Old trunks are nearly always fluted, dented or knobbly. The largest tree thus far measured in South Africa (in the Brits district) is 17 m high. The crown-spread of most of the trees measured does not exceed 17 m, and the stem diameter varies between 0,9 and 1,5 m.

Bark

The simple **leaves** are quite small, narrowly oblong or oval and are borne in decussate pairs even on thickish branchlets. The upper surface is glossy, buff-green to dark green with small, grey stipules, and the underside is brownish-green, dull and covered with silvery or brownish scales. The margin is *always rolled under* (revolute). The small, white **flowers** are borne in loose, axillary or terminal clusters (November/December).

The small **fruit** is slightly oval, and is initially green with white spots, later yellow and purplish-black when ripe. Birds and other animals eat the fruit, but it is bitter. The finely textured timber is used in the manufacturing of furniture. This tree is highly recommended as a garden plant.

Fruit

(Zimbabwe): (868) Wild olive

Olea capensis subsp. *capensis* (SA 618) **False ironwood**

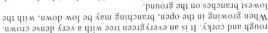

Of the six different olive trees in southern Africa (three of which are subspecies of *O. capensis*) four are totally, or for practical purposes anyway, endemic to South Africa. This particular tree is one of them and is mostly found on the coastal side of the escarpment roughly from Clanwilliam in the western Cape to the Cape Peninsula and then east and northwards to as far as northern Natal. It is abundant in both wet and dry shrub forests and the main focus of its distribution is the area between East London and Cape Town.

Fruit

Fake ironwood is one of the very variable trees. It often occurs as a bushy shrub, barely reaching 10 m in the Knysna/Tsitsikamma area, but on the slopes of Table Mountain it is a large tree. Its stem may be relatively thin, often crooked to long, thick and straight, and is fairly smooth for quite a long time. With age the **bark** becomes rough and corky. It is an evergreen tree with a very dense crown. When growing in the open, branching may be low down, with the lowest branches on the ground.

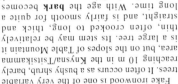

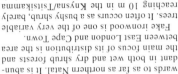

Flower buds

The **leaves** are broadly elliptic (up to 10 cm long) hard, smooth and leathery, dark green and glossy above and pale green beneath. *The leaf margin is entire and conspicuously rolled under.* The **flowers** are very small, white, sweetly-scented and are borne in many-flowered, terminal or axillary heads (July/February, depending on climatic conditions). The **fruit** is almost spherical to ovoid, up to 1 cm long, fleshy and becoming purple when ripe. The wood is widely used in the furniture industry.

Strychnos potatorum (SA 630) Black bitterberry

The various tree species in this genus differ in several respects. Some have spines and others not; some are poisonous and others not; some are poisonous and others not; while the fruits of others are edible; five have very large fruits (up to 12 cm in diameter) and the rest small (up to 2 cm). The common name of this

Fruit

tree points to the fact that it has small **fruit** (2 cm in diameter). The fruit is spherical, green at first and purplish-black when ripe. It is single-seeded and the fruit pulp has a purple. The fruit takes about a year to mature and is said to be extremely poisonous. The small, pale yellow **flowers** are stellate (star-shaped), and are borne in the axils of the new leaves (August/October). The simple **leaves** are fairly large (mostly 11 x 5 cm) and are borne in opposite pairs. The leaves are characteristic of the *Strychnos* genus, having three (*sometimes five*) *major veins originating from the base of the leaf.* Old stems are pale grey and fairly smooth, and the **bark** peels off sporadically in small, longitudinal strips.

These trees are medium-large (up to about 15 m) and single-stemmed, but the trunk invariably splits rather low down. It covers the bigger portion of Zimbabwe and is common in some regions of Mozambique as well as the Kasane/Chobe area in Botswana and the Caprivi Strip. In South Africa it is limited to the northern area of the Kruger National Park and adjoining areas to the north and north-west. A single specimen has been found along the Sabie River. (Zimbabwe: (890) Grape strychnos)

Anthocleista grandiflora (SA 632) Forest fever tree

Bark

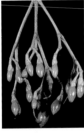

Flowers

Fruit

It may appear strange that this very large inhabitant of the northern evergreen forests in the subcontinent belongs to the same family as, among others, the *Strychnos* (Monkey-orange) and *Nuxia* (Wild elder) species. With its enormous leaves, especially in younger plants, and beautiful, fairly large, trumpet-shaped flowers, it does not seem to fit in at all.

It occurs intermittently from the northern-most point in Natal, through Swaziland, the forests in the Sabie/Tzaneen area in the Transvaal up to the Soutpansberg. Further north it is mainly restricted to the forests in the east of Zimbabwe and the Chimanimani area of Mozambique.

The Zimbabwean name for the tree, big-leaf, is quite fitting as the **leaves** may be more than 1 m long and almost 0.5 m wide. They are borne close together at the ends of the branches. The **flowers** are white, *trumpet-shaped*, sweetly scented and are borne in terminal, candelabra-like panicles. Because the trees can reach 30 m in height, the flowers are unfortunately so high up that they are rarely noticed. The green, oval **fruit** contains many small seeds and can be found almost throughout the year, as they develop very slowly.

Although this is a rather slender tree with a very long, bare trunk, it is quite decorative and should be used more often as a garden subject. Unfortunately, it will only survive in frost-free areas.

(Zimbabwe: (894) Big-leaf)

Flowers

Bark

Nuxia floribunda (SA 634) Forest elder

Four tree species represent this genus in southern Africa. This one, the forest elder is not very abundant or dominant but is a very prominent component of the plant community, especially when in flower. It occurs roughly from George in the south-eastern Cape to northern Natal and southern Mozambique, Swaziland and along the Drakensberg to the Soutpansberg and northern Lowveld. It is mainly associated with evergreen forest and therefore also inhabits the eastern area of Zimbabwe as well as the adjoining territory in Mozambique. An isolated population occurs in southern Zimbabwe.

It is evergreen and may reach a height of 15 m. Young twigs are purple and old trunks vary from fairly smooth, greyish-brown and striated to flaking or very smooth, white and with the **bark** peeling in papery strips. Stems are often crooked or forked and may have young shoots sprouting low down. The crown is large, wide-spreading and dense.

The simple **leaves** are borne in whorls of three. They are fairly large, soft, elliptic, sharply pointed, pale to dark green and glossy. The margins are entire, or slightly serrated in the upper half; the midrib may be reddish. The prominent **flowers** are fragrant, white and, although small, are *borne in masses in large, conspicuous, terminal clusters during the winter months*. Trees are sometimes covered with flowers and are then very noticeable for quite some time. The **fruit** is a tiny, ovoid capsule which splits into four segments at maturity.

The heavy, yellow wood was used by wagon builders. This worthwhile garden plant can be grown from cuttings.

(Zimbabwe: (896) Forest nuxia)

In southern Africa the toad tree occurs from northern Natal and southern Mozambique through Swaziland and the eastern Transvaal into southern Zimbabwe. It is deciduous, but only sheds its leaves late in winter and may even be semi-deciduous in wet conditions. It has a single, short trunk and, in spite of poor branching, the roundish crown is usually quite dense as a result of the large leaves. New twigs are shiny, dark green and dotted with small, corky protuberances. They occur in opposite pairs. The **bark** is soft, cork-like and splits longitudinally to form ridges that crack crosswise into oblong sections.

The simple **leaves** are borne in decussate pairs. They are oblong, very large on young plants (up to 23 x 8 cm but usually about 12 x 5 cm). They are dark green and glossy and contain latex. The sweetly-scented **flowers** are borne terminally in small panicles during spring and summer. They are trumpet-shaped with five relatively long, narrow, recurved petals. The **fruit** usually consists of two hemispheric mericarps of about 6 cm in length and 7 cm in breadth, which are joined to a common stalk. The outer layer is browny-green and dotted with pale grey tuberculate organs. At maturity the fruit dehisces while on the tree to reveal *the seeds*, which are closely packed. Each one is surrounded by a thin, bright orange, fleshy layer which is eaten by wild animals.
(Zimbabwe: (917) Toad tree)

Fruit

Dehiscent fruit

Rauvolfia caffra (SA 647) Quinine tree

Flowers

Fruit

The quinine tree favours wet conditions and is therefore mainly confined to the banks of rivers and large streams and the margins of evergreen forests. It is deciduous, in a dry habitat and evergreen, or at least semi-deciduous, in a moist habitat. It is one of the larger African trees (up to 20 m), with a single stem and fairly dense, wide-spreading crown (26 m). New twigs always occur in groups of four at the tips of the branches, but one or more die with age. The branches and stems are yellowy-grey to yellowy-brown. The old **bark** is soft and cork-like and breaks up in small pieces.

The **leaves** are simple and are borne in distinct verticils of five (seldom three or four). They are oblong, sometimes lanceolate. On young plants they are much larger than on old ones, and turn dark green and brittle when old. The midrib is conspicuously yellow-green. The small **flowers** are waxy, white, sweetly scented and are borne in large, dense cymes (September to November). The **fruit** is usually single-seeded and spherical, and about 1.5 cm in diameter. It is initially very glossy and dark green with pale dots, and when mature is dark brown to black.

(Zimbabwe: (920) Rauvolfia)

122

Members of this genus are characterized by their extraordinary and remarkable, sometimes even spectacular, flowers and it is actually a pity that only this species, with the least attractive flowers, enjoys tree status. For practical purposes this tree is South African, as it has otherwise only been found in two small localities in eastern Zimbabwe. It is associated with evergreen forests and its range extends from the eastern Cape to Natal and from there inland through Swaziland and along the escarpment roughly to Tzaneen in the eastern Transvaal.

Flowers

Although this plant usually occurs as a scrambler in thickets, it may become a small tree (up to about 4 m) when growing in the open. It then tends to form a nearly impenetrable bush with long, slender, intertwined branches. The **bark** is smooth and greenish. The simple **leaves** are arranged in whorls of three and are pendent. They are narrowly elliptic, up to 10 cm long, glossy green and ooze watery sap when damaged.

The **flowers** are striking, consisting of a bell-shaped tube of *about 1 cm long, dividing into five slender corolla lobes of about 6 cm long.* The overall colour is a dull yellow, but the broadened bases of the corolla lobes are reddish. Flowering usually occurs in September/October. The **fruit** is a paired, slender, follicular mericarp, up to 20 cm long. It becomes brown at maturity and splits longitudinally to release the seeds, which are crowned with a tuft of hair at one end.

This plant, like others in the genus, is said to be poisonous, and has apparently been used in the preparation of arrow poison. The roots, in powder form, are used for snakebites.

(Zimbabwe: (932) Poison rope)

Pachypodium namaquanum (SA 649) Elephant's trunk

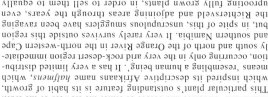
Flowers

Bark

The Apocynaceae family, of which this species is a member, is well known for the very interesting plants it harbours, mostly because of the shape, size, colour, structure and colour of either the flowers or the fruit. This particular plant's outstanding feature is its habit of growth, which inspired its descriptive Afrikaans name *halfmens*, which means 'resembling a human being'. It has a very limited distribution, occurring only in the very arid rock-desert region immediately south and north of the Orange River in the north-western Cape and southern Namibia. It very rarely survives outside this region but, in spite of this, unscrupulous smugglers have been ravaging the Richtersveld and adjoining areas through the years, even uprooting fully grown plants, in order to sell them to equally unprincipled buyers who would then try to establish them in their unsuitable gardens – obviously without success.

It rarely reaches 5 m in height, and consists of one or more cylindrical, succulent stems and no crown. The stems occasionally branch at the base or near the apex, and are thickest at the base, *tapering to the top, which is usually bent over*. A multitude of protuberances, each one spine-tipped, occur on the stem.

The single **leaves** are crowded at the top of the stem. They are simple, up to about 10 cm long, grey-green, wavy and densely velvety. The beautiful **flowers** are tubular, reddish-brown inside and green outside, up to 5 cm long and occur at the tops of the stems (August). The **fruit** occurs in twin pairs, each being fairly small, up to 4 cm long and densely covered with hairs. They dehisce while attached to the plant to release the seeds, each of which has a tuft of hair at one end.

Vitex zeyheri (SA 666) Silver pipe-stem tree

The Vitex species are characterized by digitately compound leaves, often with glands in the leaves and flowers. The sepals are usually fused to form saucer-shaped structures and two-lipped, tubiform flowers. This particular species is limited to a rather small area, extending roughly from Pretoria in a westerly direction towards Gaborone in Botswana. It favours rocky hillsides and valleys.

This is a smallish, upright, slender tree of about 5 m, with a fairly sparse crown. Young branchlets are covered with soft hairs. Old stems are dark grey and the **bark** is longitudinally fissured. The **leaves** consist of three to five leaflets which are mostly covered with soft, silvery hairs. The **flowers** are generally pale mauve, seldom cream, sweetly scented and are borne in terminal heads. The small, roughly pear-shaped **fruit** is enveloped by the enlarged, saucer-like calyx which persists on the tree for months on end. The common name refers to the fact that young twigs are often hollow, and were previously used as stems for tobacco pipes.

Flowers

Halleria lucida (SA 670) Tree-fuchsia

Flowers   *Fruit*

The tree-fuchsia is well adapted to a very wide range of climatic conditions as evidenced by its distribution as well as its occurrence in a variety of habitats, from evergreen forest to coastal and even karroid shrub. It may occasionally reach a height of 12 m but is usually encountered as a shrub, small tree, or even a climber, often with more than one trunk. Side-shoots and smaller, trailing branches usually occur from the base. The crown is sparse and fairly spreading with trailing branchlets.

The **bark** is greyish-white to pale brown, and smooth to flaking. The simple **leaves** are 4-8 cm long, rhomboidal to ovate and the margins conspicuously toothed to scalloped in the upper two-thirds only. They are borne in opposite pairs. The attractive curved, tubular **flowers** are *orange to dark red and are borne almost throughout the year* in axillary clusters, as well as on the older branches. The **fruit** is an almost spherical to ovoid, very glossy, fleshy berry which turns black when ripe. It is crowned by the long, thin, persistent style.

The wood is yellow, hard and tough but is seldom used. The edible fruit has a sweet taste and is relished by birds. This tree qualifies as a garden subject in most respects.

(Zimbabwe: (1023) Tree-fuchsia)

126

Rhigozum obovatum (SA 675) Karoo rhigozum

This species occurs mainly in the Karoo, but extends through the north-western Cape into the south of Namibia, and can also be found in the southern Orange Free State and south-western Transvaal.

The two other tree species in the genus, *R. brevispinosum* and

Flowers

R. zambesicum have a western and eastern distribution respectively, and both have spines. The first occurs in central and northern Namibia, Botswana, the northern Cape, western and north-western Transvaal and south-western Zimbabwe. The latter occurs from northern Natal, through the Lowveld of the Transvaal to southern Zimbabwe and then again in the Zambezi valley. A third species, *R. trichotomum*, which is dominant over large areas of the Kalahari, does not have tree status. All have the same yellow, very handsome, flowers.

The Karoo rhigozum is spineless, mostly multi-stemmed and may reach 3,5 m in height. The branches and twigs are rigid. Twigs are straight and may be spine-tipped. The **leaves** are trifoliolate but often simple, as only one develops. They are very small (up to 1,5 x 0,5 cm, but mostly smaller). The **Flowers** are fairly large (3 cm in diameter), *bright yellow and very striking*. They emerge only after the first good rain in spring – sometimes within days. The main flowering period is September to November and during a good season the plants are covered with Flowers. The **fruit** is a brown, flattened, narrow, smooth capsule up to 8 cm long, which dehisces longitudinally along the flat surface to release the winged seeds.

It is heavily browsed on by game.

Bark

Fruit

The bell bean tree occurs from northern Namibia in the west, through the northern areas of Botswana and Zimbabwe to Mozambique in the east, as well as the northern and north-eastern area of the Transvaal.

This species prefers well-drained soil and is therefore usually encountered on sand or rocky hillsides. It is abundant in Chobe National Park in Botswana, where some trees measure up to 12 m in height, although the majority only reach 6-7 m.

The trunk is fairly short and often branches low down, and the crown is particularly sparse and slender. The branches and stems are smooth, glossy and grey-brown to lead-grey. Very old **bark** is brownish-grey and peels off in flat, irregular flakes. The **leaves** are very long (up to 35 cm), imparipinnately compound with two to four leaflets plus a terminal leaflet. They increase markedly in size from the base to the tip of the leaf. The thin, faintly glossy leaflets on some trees are hairless, while on others they are distinctly softly hairy.

The **flowers** are bell-shaped, fairly large and beautiful. The corolla tube can be up to 5 cm long, with the lobes spreading and measuring up to 5 cm in diameter. In the Transvaal, the crinkled corolla lobes and the tube are deep maroon on the inside and pale yellow, speckled with dark red dots, on the outside. The **fruit** is unmistakable. It is a *very long (up to 60 cm), slender (1,5 cm wide), spirally twisted capsule with grey-white speckles.* (The flowers of the other species, *M. obtusifolia* are bigger and bright yellow. It has a far more limited distribution than *M. zanzibarica*, but also occurs in Chobe.)

(Zimbabwe: (103) Bean tree)

Kigelia africana (SA 678) Sausage tree

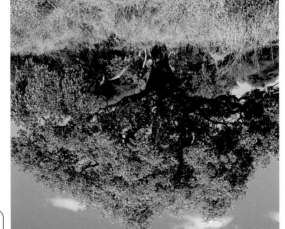

This large tree can reach 18 m in height, with an extremely thick trunk and a wide-spreading crown. It is deciduous to semi-deciduous. Old leaves are shed within a very short period of time and the new leaves, emerging shortly thereafter, are glossy and brownish-red. Old trunks are grey-brown to dark grey, and fairly smooth. The **bark** peels off in flat, irregular sections.

The **leaves** are imparipinnately compound with two to five pairs of leaflets. They are borne in verticils of three near the twig terminals, and can be up to 25 cm long. The leaflets are thin; the margins are entire, or serrated in young leaves. The large **flowers** are borne in pendent, axillary racemes of up to 50 cm long. They are roughly cup-shaped, and may become as large as 14 x 14 cm. The crumpled petals and corolla tubes are deep velvety-red on the inside, and red-brown with green, longitudinal ridges on the outside. The *sausage-shaped* **fruit** is unmistakable, measuring 50 x 10 cm or more, coloured pale brown to grey-green, and is slightly rough and very heavy. The fruit pulp is fibrous and inedible. The timber is tough, pale brown and produces a smooth finish.

(Zimbabwe: 1035) Sausage tree)

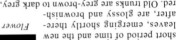

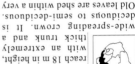

Flower

Fruit

Sesamothamnus lugardii (SA 680) Transvaal sesame bush

Two species of trees represent this genus in southern Africa.
S. benguellensis only occurs in northern Namibia; *S. lugardii* occurs from the Transvaal Lowveld (Kruger National Park) through the northern Transvaal to the south of Zimbabwe and the east of Botswana ('Tuli Block'). All members of the Pedaliaceae family, to which this tree belongs, are characterized by large, beautiful trumpet-shaped flowers.

Flowers

The Transvaal sesame bush is a small, deciduous tree, measuring 4–5 m, with a very sparse, poorly branched crown but an abnormally thick, succulent-like trunk which always subdivides near the ground. The **bark** on older branches and trunks is yellowish-brown and fairly glossy. The thin, outer layer of bark peels off to expose the green, living bark. Single thorns are set spirally on the twigs; initially they are soft with leaf-like appendages, but later become hard and sharp. The **leaves** are rather small and are borne in groups above the thorns, even on thick branches. The **flowers** of this species are unique, but unfortunately very few are produced. Each consists of a particularly long, thin, red-brown corolla tube which is elongated at the base into what is known as a 'spur', which extends beyond the flower's junction with the twig. The petals are crinkled and white to pale cream-coloured. The **fruit** is a flat capsule of approximately 6 x 5 cm, which resembles that of the jacaranda. The flat seeds are winged. The timber is soft, fibrous and worthless.

Bark

(Zimbabwe: (1036) Sesame bush)

130

Breonadia salicina (SA 684) **Matumi**

Bark

Flowers

The only specimen on which statistics have been published by the Dendrological Society of South Africa grows in the Transvaal (Letaba district): height 41 m; spread 23,6 m; diameter of the trunk at breast height 2,06 m. The average tree occurring in the Transvaal Lowveld has a *fairly short stem and a densely foliaged, poorly spreading crown, and often branches as low as ground level.* Old stems are grey-brown, and although the **bark** breaks up into irregular ridges, it does not peel off.

Matumi is a fairly fast-growing evergreen that occurs from northern Natal through Swaziland and the eastern parts of the Transvaal to the western side of Mozambique, and through eastern Zimbabwe up to the Zambezi River. North of that it spreads out across Africa.

The simple **leaves** are borne in verticils of four and are usually set close together at the twig terminals. They are long and narrow (up to 25 x 4 cm), dark green, glabrous, smooth and particularly glossy. The minute **flowers** are pale yellow with a pink tinge, and are borne in dense, spherical heads of 2 cm in diameter from November to March. They are borne singly in the leaf-axils. The small **fruit** is inconspicuous.

The outstanding asset of this tree is its timber, the quality and availability of which nearly led to its exploitation. Large trees were felled at such a rate in the Transvaal Lowveld that the authorities had to intervene, and it is now a protected species in South Africa. The wood is heavy, fairly hard, oily, pale to dark brown and frequently blotched.

(Zimbabwe: (1057) Matumi)

131

Bark

Fruit

Flowers

The wild pomegranate mainly inhabits the evergreen forests at the coast and the mountains separating the low-lying southern and eastern coastal belt from the inland plateau of South Africa. Its range extends from near Swellendam in the Cape to roughly as far north as Tzaneen in the Transvaal. It grows in dry forest, grassland, open woodland and even in swampy terrain.

This is an evergreen shrub or small tree which usually does not exceed 6 m in height. Even in the open it is upright and slender. The **bark** is greyish-brown and smooth. It usually has a rather crooked trunk. The large, very dark green **leaves** are simple and are borne in opposite pairs. *Transverse ridges at the nodes, between the leaves of a pair*, are characteristic. The outstanding asset of this tree is its **flowers** which are most attractive and showy. *They are orange to crimson-red, tubular*, up to 2,5 cm long and are borne in dense clusters at the twig terminals. Copious amounts of nectar, which has a strong, sweet scent, are produced. According to various authors, flowering mainly occurs in spring and summer, but those in the photograph were found in Kirstenbosch during early August. The very attractive **fruit** is urn-shaped and occurs in dense clusters. Leathery, horny, persistent calyx-lobes crown the fruit.

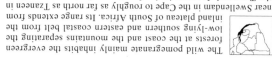

Burchellia bubalina (SA 688) Wild pomegranate

132

Seven *Gardenia* species, which are remarkable trees, endowed with *beautiful, white, sweetly-scented flowers*, occur in southern Africa. The best-known is probably *G. volkensii*, which is widespread in the Transvaal, and northern Natal, Zimbabwe, Mozambique, Botswana and Namibia.

The Natal gardenia inhabits a relatively small area only, occurring in the northern part of the Natal coastal region, and probably also in Swaziland and southern Mozambique. It is a small, densely-branched tree. The trunk and branches are pale grey, very smooth and hard and the **bark** flakes occasionally. It reaches only about 4–5 m in height.

The **leaves** are simple and are crowded near the ends of the short, stiff, nearly spinescent branchlets. They may measure up to 5 x 2,5 cm, and are a glossy light green in colour, and wavy. The beautiful **Flowers** are white, aging to yellow, and are sweetly scented. The corolla tube can be up to 6 cm long, and the spreading lobes about 3 cm long. The **fruit** is distinctive, ovate, smooth, up to 5 cm long, and bright orange-yellow. The persistent calyx lobes initially crown the apex, but later wither away. The fruit remains on the trees for an extended period of time.

Although the wood is fine-grained and hard, and should therefore be suitable for the manufacturing of smaller articles, sufficiently large pieces are not obtainable. It is used as fire-wood.

Flowers

Fruit

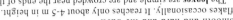

Gardenia cornuta (SA 690,1) Natal gardenia

133

Gardenia volkensii subsp. *volkensii* (SA 691) Transvaal gardenia

This is the most common and widespread of the genus, featuring in northern Namibia and northern Botswana through the western half of Zimbabwe and eastern Botswana, nearly all the bushveld areas of the Transvaal and southern Mozambique to northern Natal and Swaziland. It occurs on a wide variety of soil types, ranging from well-drained sand to poorly drained, brackish, clayey soil, as well as rocky situations. This is one of the smaller tree species, seldom reaching 7–8 m. The trunk is always *short, relatively thick and often fluted.* The smooth **bark** is pale grey with a yellow tinge. It flakes in small, fairly thick sections, resulting in a mottled appearance. Twigs are borne in whorls of three and are very hard and stiff. The crown spreads wide and is relatively dense. The leaves are occasionally shed in winter.

Flowers

The spatulate (spoon-shaped) **leaves** are borne in whorls of three. The beautiful, white, waxy **flowers** turn yellow with age, and may be up to 10 cm in diameter. Flowering takes place between September and December, possibly later. The **fruit** is mostly ovate, but may be nearly spherical. They are grey, prominently ribbed longitudinally, and covered with greyish-white encrustations. They remain on the trees for an extended period, and fall unopened. Only the wood is yellowish, very hard, heavy and fine-grained. Only a lack of large pieces prevents it from being widely used. This tree is an outstanding garden subject.

(Zimbabwe: 1074) Common gardenia

Rothmannia capensis (SA 693) Cape Gardenia

Although this tree may reach as much as 20 m in height in forest situations, it usually does not exceed about 9 m. It occurs over a wide range of altitudes from sea level to about 1 600 m, and from evergreen forest and riverine vegetation to rocky hillsides. It is predominantly a South African species, extending from the south-western Cape, all along the coast to Natal and Swaziland and then to the Transvaal, along the Drakensberg to the north but also on the mountains to the west as far as Zeerust and the adjoining area in Botswana.

The Cape gardenia is a fairly slender, evergreen tree with a single trunk, frequently branching low down. The crown is moderately dense and only spreads in favourable habitats. Young branches are smooth and brown. The **bark** on old stems is fairly smooth and dark grey, mottled with brown. The large **leaves** are a glossy dark green and leathery, and are borne close together at the twig terminals. They are characterized by small but prominent swellings *in some of the vein-axils*.

The sweetly-scented, attractive **Flowers** are widely trumpet-*shaped with creamy-white petals and reddish markings in the throat* (December/January). The **fruit** is oval, smooth, shiny and measures up to 7 cm long. It has prominent longitudinal grooves as well as a circular marking at the apex, and will remain on the tree for nearly a full year.

Flower

Bark

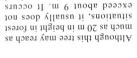

Fruit

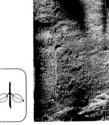

Rothmannia fischeri subsp. fischeri
(SA 694) Woodland rothmannia

Flowers

Bark

This species' distribution range extends from Maputaland (northern Natal) in a narrow belt through Mozambique up to the Limpopo where it enters South Africa. It then spreads north again into Zimbabwe where it occurs in a narrow strip parallel to Mozambique, extending about as far north as Harare, then spreading east again to the Indian Ocean.

Fruit

It is a small tree (up to 8 m), very slender, evergreen with a long, bare trunk and a fairly dense crown, frequenting a rocky, well-drained habitat. On older branches the **bark** is distinctly reddish-brown. The colour is retained on old stems but is only revealed when the outer layer of dead bark peels off in small, flat strips.

The fairly large, glossy, dark green **leaves** are up to 9 cm in length and are borne in opposite pairs. Small pockets occur in the axils of the veins on the under surface, in the middle section of the leaf only. The **flowers** are showy, funnel-shaped, terminal and solitary. The funnel terminates in five tapering petals arranged in a star-pattern. The exterior of the tube is pale green, the petals are white with red speckles and the interior of the tube is greenish-white with the same speckles. The **fruit** is oval, shiny, dark green and up to 7,5 cm long. It has a conspicuous grey, circular mark at the apex.

(Zimbabwe: (1075) Woodland rothmannia)

Alberta magna (SA 701) Natal flame bush

KRISTO PIENAAR

When seeing this tree in flower for the first time, the first thought which comes to mind is that it cannot be indigenous. However, its beautiful, exotic appearance is probably due to the fact that it is an inhabitant of the ravines and evergreen forests in the Transkei and Natal.

Although encountered most often as a large shrub, this evergreen, medium-sized tree may reach a height of more than 10 m. It is multi-stemmed or low-branching. The **bark** on young trunks is a greyish-brown and fairly smooth but with transverse grooves; it becomes rough with age. The crown of the tree is dense and fairly wide-spreading.

The **leaves** are dark green and glossy above and a paler green on the underside. They are elliptic and are borne in opposite pairs. The midrib and secondary veins are yellowish and conspicuous, particularly on the underside.

The dramatically striking, tubular **flowers are dark red to crimson** and are borne in dense axillary or terminal heads. Palgrave (1983) claims the flowering time to be from January to April, but as late as August, I have encountered a cultivated tree growing in Jonkershoek in the Cape bearing both fruit and flowers. The ovate **fruit** is small (0.5 cm long) but is particularly conspicuous due to the two brilliant red, membranous wings that are attached to its apex (February/August).

This spectacular tree is a must for all gardens with a suitable climate, though it is usually slow-growing. Unfortunately, it would probably only thrive in warm and humid conditions. It can be grown from cuttings.

Flowers

Fruiting stage

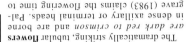

137

Vanguerria infausta (SA 702) Wild medlar

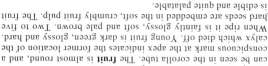

Of the four wild medlar species occurring in southern Africa, this one is by far the most widespread and common, extending from the north-eastern Cape and Transkei northwards in a horse-shoe pattern into Mozambique and the Transvaal, nearly blanketing the latter area. It also occupies most of Zimbabwe, the east and north-east of Botswana and the major part of northern Namibia. It seems to prefer a well-drained substrate such as sandy soil, and often grows on rocky hillsides.

It is a small, deciduous tree of 6 m, with a short trunk and a dense, spreading crown with branches usually as low as ground level. Branches and stems have pale grey-brown **bark,** sometimes with a yellow tinge. The thin outer layer of bark peels off in flat, untidy strips.

The simple **leaves** are borne in opposite pairs, set fairly far apart. They are quite large, usually measuring about 11 x 6 cm, but sometimes twice as big, very soft and slightly pubescent. The small, yellow-green **flowers** are borne in dense racemes in the leaf-axis or immediately above the scars left by the fallen leaves. Conspicuous hairs can be seen in the corolla tube. The **fruit** is almost round, and a conspicuous mark at the apex indicates the former location of the calyx which died off. Young fruit is dark green, glossy and hard. When ripe it is faintly glossy, soft and pale brown. Two to five hard seeds are embedded in the soft, crumbly fruit pulp. The fruit is edible and quite palatable. (Zimbabwe: (1096) Wild medlar)

Fruit

Flowers

Chrysanthemoides monilifera (SA 736,1) Bush-tick berry

Although this plant is very seldom seen in its true tree form of
about 6 m in height, *its sheer
beauty when in flower*, especial-
ly as it occurs in winter, merits its inclusion
in this guide. Furthermore, it inhabits a
large area of South Africa, occurring in a
broad belt from roughly Springbok in
Namaqualand in the west, following the
coast line around the Cape to northern
Natal, and then along the escarpment to the
Soutpansberg. Unlike the vast majority of
other tree species, it can tolerate even the
very low temperatures prevailing in Lesotho

Flowers

and the eastern Orange Free State during winter. Its abundance on
coastal dunes, often near the high-tide mark, further emphasizes its
remarkable adaptability.

It is evergreen and is usually encountered as a very dense, multi-
stemmed, large shrub with a roundish crown. Its **bark** is fairly
smooth and grey. The simple **leaves** are borne alternately; they are
rather large (up to 7 cm) and are covered with a dense, white layer
of hairs when young but become glossy with age. The leaf margins
are conspicuously toothed in the upper half to one-third only.

Bush-tick berry belongs to the sunflower family (Asteraceae)
and the **flowers** are daisy-like, as can be seen in the picture. They
are bright yellow, fairly large (4 cm in diameter), and are borne at
the ends of the branches, either solitary or in small groups (May/
October). The species name, which means 'bearing a necklace',
aptly describes the **fruit** and its configuration. It is almost spherical, about 0,6 cm in diameter, glossy, fleshy and arranged in close
proximity around the edge of the receptacle. The fruit becomes
purple at maturity.

(Zimbabwe: (1172) Bush-tick berry)

139

Glossary

alternate – of a leaf or flower arranged singly at different heights on either side of the stem

aril – an outer covering of some seeds, often brightly coloured

axil – angle between the petiole of a leaf and a twig

bipinnate – having leaflets growing in pairs on paired stems

conduplicate – one half of a leaf is folded lengthwise and upwards upon the other

corolla – petals of a flower

cyme – an often flat-topped inflorescence which blooms from the centre outwards, and whose main axis always ends in a flower

deciduous – shedding leaves once a year

decussate – succeeding pairs of leaves crossing at right angles

dehiscent – splitting open to release seeds

drupe – one-celled fruit with one or two seeds, e.g. a plum

edaphic – relating to the physical and chemical properties of soil

elliptic – oval-shaped

entire – a continuous, unimpaired leaf margin

floret – a small flower of a composite flower

follicle – single-chambered fruit that splits only along one seam to release its seeds

genus (genera) – group of closely related species

glabrous – with an even, smooth surface; hairless

heartwood – usually darker, harder, central part of a trunk

imparipinnate – pinnate leaf with an odd terminal leaflet

indehiscent – a fruit which does not split to release the seeds

inflorescence – a number of flowers on a common stalk

lanceolate – lance-shaped

latex – the fluid (often sticky) found in some plants

mericarp – a fruit that develops from one of the carpels of the ovary

midrib – large, central vein in a leaf

oblanceolate – inversely lanceolate

obovate – inversely egg-shaped

orbicular – round or shield-shaped

ovate – egg-shaped

palmate – of four or more leaves arising from a common stalk

panicle – tuft or bunch of flowers, close or scattered

paripinnate – pinnate leaf without a terminal leaflet

peduncle – stem or stalk supporting flower or fruit

pendent – hanging down

pergamentaceous – parchment-like

pericarp – the part of a fruit enclosing the seeds

pinna – primary division of a pinnate leaf (leaflet)

pinnate – compound leaf divided in a feathery manner

pubescent – covered with soft hairs or down

raceme – inflorescence in which the flowers are borne along the main stem, with the oldest flowers at the base, e.g. hyacinth

rachis – stalk or axis of a compound leaf bearing the pinnae

recurved – rolled or bent backwards

sapwood – outer layer of wood between the bark and the heartwood

savanna – subtropical or tropical grassland with scattered trees and shrubs

serrate – of a leaf margin notched like the teeth of a saw

sessile – of a leaf or flower sitting directly on a base without a supporting stalk, petiole, etc.

spathe – a large bract enclosing the inflorescence

spicate – having, or arranged in, spikes

spike – inflorescence with sessile flowers along a central axis

spinescent – spiny, having sharply pointed thorns
stipules – small, leaf-like appendages which occur in pairs at the junction of the leafstalks with a twig, mostly falling early
terminal – situated at the tip
tomentose – covered with matted hairs
trifoliolate – having three leaves or leaflets
trilocular – having three chambers or cavities
tuberculate – covered with tubercles (protuberances)
tubiform – tube-shaped
umbel – flowers forming a cluster from a common centre
verticil – a circular arrangement of leaves around a stem
whorl – of three or more leaves arising in a circle at the same point on a twig

Flower

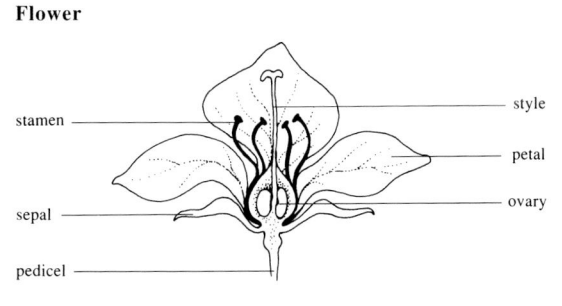

Compound leaf

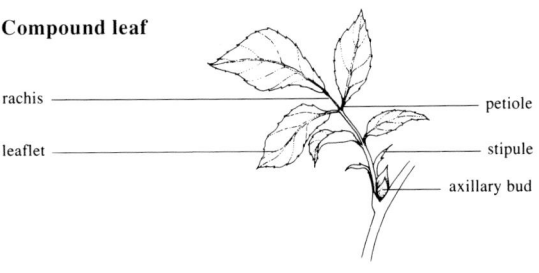

Simple leaf

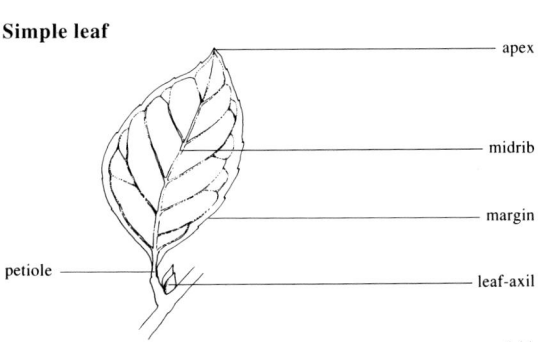

Index

Acacia erioloba 32
Acacia hebeclada 33
Acacia karroo 34
Acacia tortilis 35
Acacia xanthophloea 36
Adansonia digitata 88
Adenolobus garipensis 46
African mangosteen 93
African star-chestnut 90
Afzelia quanzensis 44
Alberta magna 137
Albizia adianthifolia 28
Albizia harveyi 29
Albizia versicolor 30
Aloe dichotoma 11
Ana tree 31
Anthocleista grandiflora 119
Apple-leaf lance-pod 58
Apple-leaf thorn-tree 31
Azanza 87
Azanza garckeana 87
Baikiaea plurijuga 43
Balanites pedicellaris 62
Baobab 88
Baphia massaiensis 54
Bauhinia petersiana 47
Bead-bean 26
Bean tree 128
Beechwood 17
Big-leaf 119
Black bitter-berry 118
Bladder-nut 114
Bloodwood 57
Blossom tree 52
Bolusanthus speciosus 53
Boscia albitrunca 25
Brabejum stellatifolium 16

Brachystegia boehmii 41
Breonadia salicina 131
Burchellia bubalina 132
Burkea 38
Burkea africana 38
Burning-bush combretum 101
Bush-tick berry 139
Butterfly leaf 46
Calodendrum capense 63
Camel's foot 48
Camel thorn 32
Candle thorn 33
Candle-pod acacia 33
Cape ash 69
Cape beech 108
Cape chestnut 63
Cape gardenia 135
Cape plum 21
Cape stock-rose 86
Carrot tree 107
Cassia abbreviata 49
Cassine papillosa 82
Chrysanthemoides monilifera 139
Coastal red-milkwood 111
Colophospermum mopane 39
Colpoon compressum 21
Combretum apiculatum 99
Combretum imberbe 100
Combretum microphyllum 101
Commiphora marlothii 66
Commiphora mossambicensis 67
Common cluster fig 14
Common coral tree 61
Common false-thorn 29
Common gardenia 134
Common resin tree 79

Common saffron 82
Common saffronwood 82
Common star-chestnut 91
Common sugar-bush 19
Common tree euphorbia 74
Common wild pear 89
Cordyla africana 51
Cork bush 56
Cross-berry 85
Croton megalobotrys 72
Cussonia natalensis 106
Dais cotinifolia 98
Diospyros mespiliformis 113
Diospyros whyteana 114
Dog plum 69
Dombeya rotundifolia 89
Cassine
Ebony
Ebony diospyros 113
Ebony tree 112
Ekebergia capensis 69
Elephant's trunk 124
Englerophytum magalismontanum 110
Entandrophragma caudatum 68
Erythrina lysistemon 61
Euclea pseudebenus 112
Euphorbia ingens 74
Faidherbia albida 31
False ironwood 117
Faurea saligna 17
Fever tree 36
Fever-berry croton 72
Ficus salicifolia 13
Ficus sycomorus 14
Ficus verruculosa 15

Flame creeper 101
Flat-crown 28
Forest elder 120
Forest fever tree 119
Forest nuxia 120
Fountain bush 55
Fried-egg flower 95
*Friesodielsia
obovata* 23

*Garcinia
livingstonei* 93
*Gardenia
cornuta* 133
*Gardenia
volkensii* 134
Grape strychnos 118
Green-apple 22
*Grewia
occidentalis* 85
*Guibourtia
coleosperma* 40
*Gyrocarpus
americanus* 24

Halleria lucida 126
*Harpephyllum
caffrum* 76
Hottentot's
cherry 83
*Hyphaene
coriacea* 10

Indaba tree 84

Jackal-berry 113
Jacket plum 84
Jasmine pea 54
*Julbernardia
globiflora* 45

Kalahari apple-
leaf 58
Karoo boer-bean 42
Karoo
rhigozum 127
Karree 81
*Kigelia
africana* 129
*Kiggelaria
africana* 96
Kirkia wilmsii 65

Lala palm 10
Large false
mopane 40
Large fever-
berry 72
Large-leaved
false-thorn 30

Leadwood 100
Lebombo wattle 37
Lesser torch-
wood 62
*Leucospermum
conocarpoden-
dron* 18
*Lonchocarpus
nelsii* 58
Long-tail cassia 49
Lowveld
mangosteen 93
Lowveld
newtonia 37
Loxostylis alata 78
Lucky-bean tree 61

*Maerua
angolensis* 26
Manica protea 19
Manketti tree 73
*Markhamia
acuminata* 128
Marula 75
Matumi 131
*Maurocenia
frangularia* 83
*Mimusops
caffra* 111
Mobola plum 27
Monkeybread 48
*Monodora
junodii* 22
Monotes glaber 94
Mopane 39
Mountain cedar 9
Mountain
cypress 9
Mountain hard
pear 97
Mountain
mahogany 68
Mountain
seringa 65
*Mundulea
sericea* 56
Munondo 45

Natal flame
bush 137
Natal gardenia 133
Natal mahogany 70
Natal wild
banana 12
*Newtonia
hildebrandtii* 37
Northern dwaba-
berry 23
*Nuxia
floribunda* 120

Nyala tree 60

Ochna pulchra 92
Olea capensis 117
Olea europaea 116
*Olinia
emarginata* 97
Oncoba spinosa 95
Outeniqua
yellowwood 8
*Ozoroa
paniculosa* 79

*Pachypodium
namaquanum* 124
Pale-fruited
monotes 94
Paperbark
commiphora 66
Pappea capensis 84
*Parinari
curatellifolia* 27
Peeling plane 92
Peeling-bark
ochna 92
*Peltophorum
africanum* 50
Pepper-leaved
commiphora 67
*Piliostigma
thonningii* 48
Pink wood 96
Pod mahogany 44
*Podocarpus
falcatus* 8
Poison rope 123
Poison-pod
albizia 30
Pompon tree 98
Prince-of-Wales'
feathers 41
Propeller tree 24
Protea caffra 19
Protea mundii 20
*Protorhus
longifolia* 77
*Psoralea
pinnata* 55
*Pteleopsis
myrtifolia* 102
*Pterocarpus
angolensis* 57

Quinine tree 122
Quiver tree 11

Rapanea 108
*Rapanea
melanophloeos*
108

Rauvolfia 122
Rauvolfia
caffra 122
Red beech 77
Red bushwillow 99
Red currant
rhus 80
Red seringa 38
Resin tree 79
Rhigozum
obovatum 127
Rhodesian teak 43
Rhus
chirindensis 80
Rhus lancea 81
Rock cabbage
tree 106
Rothmannia
capensis 135
Rothmannia
fischeri 136
Rough-bark flat-
crown 28

Sand camwood 54
Sausage tree 129
Savanna dwaba-
berry 23
Schinoziophyton
rautanenii 73
Schotia afra 42
Schrebera alata 115
Sclerocarya
birrea 75
Securidaca
longipeduncu-
lata 71
Sesame bush 130
Sesamothamnus
lugardii 130
Shepherd's
tree 25
Sickle-leaved
albizia 29
Sideroxylon
inerme 109
Silver pipe-stem
tree 125
Silver
terminalia 104
Simple-leaved
cabbage tree 106
Sjambok pod 49

Small green-
thorn 62
Small-leaved
terminalia 103
Snot apple 87
Snuff-box tree 95
Southern ilala
palm 10
Sparrmannia
africana 86
Squat sterculia 91
Steganotaenia
araliacea 107
Stem-fruit 110
Sterculia
africana 90
Sterculia
rogersii 91
Stink bush-
willow 102
Strelitzia
nicolai 12
Strophanthus
speciosus 123
Strychnos
potatorum 118
Sweet thorn 34
Sycamore fig 14
Syzygium
guineense 105

Tabernaemontana
elegans 121
Tarwood 78
Terminalia
randii 103
Terminalia
sericea 104
Toad tree 121
Transvaal beech 17
Transvaal
gardenia 134
Transvaal
milkplum 110
Transvaal sesame
bush 130
Tree pincushion 18
Tree wistaria 53
Tree-fuchsia 126
Trichilia
emetica 70
Two-winged
pteleopsis 102

Umbrella thorn 35

Vangueria
infausta 138
Vepris
lanceolata 64
Violet tree 71
Virgilia
oroboides 52
Vitex zeyheri 125

Water fig 15
Water pear 105
Weeping wattle 50
White bauhinia 47
White
ironwood 64
White
milkwood 109
White
sugarbush 20
Widdringtonia
nodiflora 9
Wild almond 16
Wild jasmine 115
Wild mango 51
Wild medlar 138
Wild olive 116
Wild peach 96
Wild pear 89
Wild plum 76
Wild pome-
granate 132
Wild strelitzia 12
Wild teak 57
Willow rhus 81
Wing pod 59
Wing-leaved
wooden pear 115
Wonderboom fig 13
Wooden
banana 68
Woodland
rothmannia 136
Woodland
waterberry 105

Xanthocercis
zambesiaca 60
Xeroderris
stuhlmannii 59

Zambezi teak 43